PREFACE

Our aim with this edition, as with the others, is to cover comprehensively the techniques of modern plastering and to emphasise basic practical craft skills. The encyclopaedic form will, we hope, enable the plasterer to select information for a specific job at whatever stage it is needed; thus, a student may start with basic fundamentals and a more experienced craftsman, faced with an unfamiliar task, may acquaint himself with the finer points.

All the techniques of fibrous plastering are dealt with comprehensively and solid work is described in full. The principles of glass-fibre work are outlined. Methods now obsolete are detailed to provide a craftsman with the knowledge necessary to undertake restoration and reproduction. The book also provides mathematical formulae, a comprehensive glossary of terms, and the geometry necessary for setting out practical work.

We have found that architects and designers are often not fully aware of the possibilities of plaster-work and we hope to shed new light on the scope of the craft and inspire its fuller exploitation in contemporary building practice.

The task of preparing this work was a formidable one, and we decided early on in the venture that the most practical way of tackling it was for each of us to be responsible for certain parts of the book. Don Stagg took responsibility for the entries relating to solid plastering; Brian Pegg for the fibrous work.

This third edition of *Plastering* has been thoroughly revised by Brian Pegg and details are given of new manufacturers' materials and processes. An Appendix has been devoted to health and safety to take account of all the legislation that has been enacted in recent years and the increasing emphasis on safety at work. With this in mind too, the section on scaffolding has been enlarged and updated, as has the list of British Standards.

Brian Pegg
Don Stagg

v

LIST OF ILLUSTRATIONS

Plates 1–4, between pages 262 and 263, show tools used in plastering.

List of Illustrations

List of Illustrations

LIST OF RELEVANT BRITISH STANDARDS AND CODES OF PRACTICE

British Standards and British Standard Codes of
Practice may be obtained from the British Standards
Institution, 2 Park Street, London W1A 2BS.

BS 12 : 1996	Portland cement
BS 410 : 1986 (1995)	Test sieves
BS 476 (Many parts)	Fire tests on building materials and structures
BS 812 (Many parts)	Testing aggregates
BS 882 : 1992	Aggregates from natural sources for concrete
BS 890 : 1995	Building limes
BS 1014 : 1975 (1992)	Pigments for Portland cement and Portland cement products
BS 1191 : Part 1 : 1973 (1994)	Gypsum building plasters. Excluding premixed lightweight plasters
BS 1191 : Part 2 : 1973 (1994)	Gypsum building plasters. Premixed lightweight plasters
BS 1192 : Part 3 : 1987 (1997)	Symbols and other graphic conventions
BS 1192 : PG : 1988	Graphic symbols for construction drawings (Pocket Guide)
BS 1199 & 1200 : 1976 (1996)	Building sands from natural sources
BS 1217 : 1997	Cast stone
BS 1230 : Part 1 : 1985 (1994)	Gypsum plasterboard excluding materials submitted to secondary operations
BS 1369 : Part 1 : 1987 (1994)	Expanded metal and ribbed lathing
BS 3148 : 1980	Tests for water for making concrete
BS 3496 : 1989 (1995)	E glass fibre chopped strand mat for the reinforcement of polyester and other liquid laminating systems
BS 4022 : 1970 (1994)	Prefabricated gypsum wallboard panels
BS 4027 : 1996	Sulfate-resisting Portland cement
BS 4131 : 1973	Terrazzo tiles
BS 4357 : 1968	Precast terrazzo units
BS 4549 : Part 1 : 1997	Guide to the preparation of a scheme to control the quality of glass reinforced polyester mouldings
BS 4550 : Six parts	Methods of testing cement
BS 4551 : Part 1 : 1998	Physical testing of mortars, screeds and plasters
BS 4551 : Part 2 : 1998	Chemical and aggregate grading

BS 4721 : 1981 (1986)	Ready-mixed building mortars
BS 4887 : Part 1 : 1986	Mortar admixtures. Air-entraining (plasticizing) admixtures
BS 4887 : Part 2 : 1987	Mortar admixtures. Set-retarding admixtures
BS 5262 : 1991	Code of practice for external renderings
BS 5270 : Part 1 : 1989 (1997)	Polyvinyl acetate (PVAC) emulsion bonding agents for indoor use with gypsum building plasters
BS 5385 : Five parts	Wall and floor tiling
BS 5492 : 1990	Code of practice for internal plastering
BS 5973 : 1993	Code of practice for access and working scaffolds and special scaffold structures in steel
BS 5974 : 1990	Code of practice for temporarily installed suspended scaffolds and access equipment
BS 6100 : Part 6 (Six sections)	Glossary of building terms. Concrete and plaster
BS 6452 : Part 1 : 1984 (1994)	Beads for internal plastering and dry lining. Galvanised steel beads
BS 6477 : 1992	Water repellents for masonry surfaces
BS 8000	Workmanship on building sites
BS 8000 : Part 4 : 1989	Code of practice for waterproofing
BS 8000 : Part 8 : 1994	Code of practice for plasterboard partitions and dry linings
BS 8000 : Part 9 : 1989	Code of practice for cement/sand floor screeds and concrete floor toppings
BS 8000 : Part 10 : 1995	Code of practice for plastering and rendering
BS 8204 : Five parts	Screeds, bases and in-situ floorings
BS 8212 : 1995	Code of practice for dry lining and partitioning using gypsum plasterboard
BS 8290 : Three parts	Suspended ceilings

Note: Many British Standards are being replaced by harmonised European Standards with the designation 'BS EN'.

BS EN 133 : 1991	Respiratory protective devices: classification
BS EN 149 : 1992	Filtering half masks to protect against particles
BS EN 166 : 1996	Personal eye protection
BS EN 346 : Two parts	Protective footwear for professional use
BS EN 397 : 1995	Industrial safety helmets

THE ENCYCLOPAEDIA

ABACUS

The top member of a classical capital. It is usually square on plan but may have concave sides. A reverse mould is made by running a moulding and cutting and fitting it to the required shape.

ACCELERATOR

An ADMIXTURE that speeds up the set, hardening or curing of a mix. For plasters alum, for cements calcium chloride, for polyester resins see GLASS FIBRE. See also Appendix p. 278.

ACETONE

See Appendix p. 278.

ADDITIVE

Material, other than aggregates, cement, plaster or lime, added to the binder to modify the properties of the mix or the finished product. Accelerators, waterproofing compounds, colours, plasticisers and hardeners (all in powder form) are available for adding to cement.

ADHESION

The bond between a material in plastic form and the surface to which it is applied. See also BACKGROUND, bonding agents.

ADHESIVE

For sticking plaster to plaster: neat POLYVINYL ACETATE or burnt SHELLAC. See also BACKGROUND, bonding agents.

ADMIXTURE

Material, other than aggregates, cement, plaster or lime, added to the mix to modify the properties of the mix or the finished product. Admixtures *to cement*-adhesive, waterproofer, floor-laying admixtures, plasticiser, frost proofer, hardener/dust proofer, accelerator, retarder; *to plaster*-accelerator, hardener, retarder.

AGGREGATE

Inert filler added to plasters and cements to provide bulk and produce some desired effect.

Barytes

Used in X-ray plaster. A radiation-resistant aggregate produced from barium sulphate.

Granite chippings

Used to form pavings, stairs, etc., for heavy wear. The size of the aggregate can vary from 3 mm to 10 mm. 6 mm is most commonly used in plastering but 10 mm can be used for large floor areas or externally. Granite should contain a percentage of dust to assist with its workability and produce a better finish. It is usually limited to 10% of the total bulk. See also FLOOR LAYING AND ROOF SCREEDING; STAIRCASE WORK IN GRANITE AND CEMENT

Gravel

Small stones formed by the natural disintegration of rock. Pea gravel (even, round, small pebbles) is suitable for pebble dashing. See FINISH/RENDERING, EXTERNAL.

Marble chippings

Used in TERRAZZO flooring. Must be sharp and angular and free from fines and impurities.

Pebbles or shingle

Small, round stones, suitable for pebble dashing.

Perlite

Produced from perlite ore, a crude natural glass containing chemically combined water and having a concentric shelly structure. It is crushed and graded, pre-heated to soften the glass and then suddenly subjected to a high temperature. This causes the water to turn into steam and blow the material into glass bubbles. Used as a lightweight aggregate.

Pumice

Used in ACOUSTIC PLASTERING.

Sand

See SAND.

Spar

Clean white or pink stone chippings, used in pebble dashing.

Stone, crushed

Coarse aggregate. Suitable for rough casting and terrazzo.

Vermiculite

Mica. A laminated material with one million laminations to the inch. It is heated until its chemically combined water turns to steam and pushes the layers apart like a concertina. This process is known as exfoliation. Used as a lightweight aggregate.

AIR HOLES

1 On faces of casts – see CASTING IN FIBROUS PLASTER.
2 Drilled in cases – see MOULD, REVERSE, CLAY-CASE and RUN-CASE.

ALABASTER

The purest, finest form of gypsum. Translucent and marble-like in appearance.

ALUM

Potassium aluminium sulphate, used in plastering either as an accelerator for gypsum plaster, or for treating the surface of GELATINE moulds before casting.

In either case the alum should be dissolved in warm water; as an accelerator it is then added to the gauging water for plaster. The plaster is then added and the action of the alum speeds up the act of crystallisation and effects an early set. It tends to weaken the final hardness of the set plaster.

ANGLE, EXTERNAL FORMED, INTERNAL PLASTERING

There are three types – Fig 1 (a) *pencil round*, (b) *bullnose* and (c) *chamfer* – and are all formed by hand without the assistance of a running mould. However, to ensure an accurate angle the bullnose and the chamfer (or splayed angle) purpose-made templates, as illustrated, are used to form. The term 'pencil round' is rather a loose one and covers anything from a sharp angle to one the approximate diameter of an ordinary pencil. A sharp angle may be formed merely by passing a wet crossgrain float up

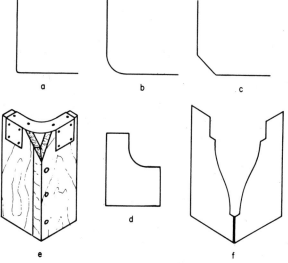

Fig. 1. Angle (solid): templates and gauges

and down the finished angle as it hardens and then using an external-angle trowel, or finishing by hand.

The bullnose consists of a plain quarter round, the radius being normally from 12.5 mm upwards. A plain angle template (d) is sufficient to give the correct shape to the floating while a box template (e) can be used, for the finishing. However, once the correct shape is formed in the floating, many plasterers can produce a near-perfect finished angle merely with the laying trowel.

The system for forming a plain chamfer is the same. A splayed-angle template is used for the floating coat and a box template for the finish. This will be largely dependent upon the size of the chamfer, as many craftsmen prefer to fix rules at both sides of the angle and work across the two rules.

In all cases the craftsman's judgement should be his guide as to which method to use. Box templates can take the size of the angle from that produced by the external-angle trowel up to a maximum radius or splay of about 75 mm. After this, larger gauges and templates should be purpose made.

A bird's beak gauge (f) or metal template will normally be required to bring the angles to a sharp finish just above the skirting and below the cornice or ceiling.

ANGLE, EXTERNAL RUN, INTERNAL PLASTERING

The three basic types of angle bead are the ovolo, the staff bead and the moulded chamfer. All will require a running mould and for the best possible results they should be run on a slipper rule and a nib rule. The floating must be completed before the running is commenced. Also, any angle should be cut away so that the running mould is not fouled. One of two methods may be used when running, using nib and slipper screeds of lime-putty plaster or, alternatively, running on a rebate.

To run on two rules the slipper rule must be fixed first. This is plumbed down and fixed so that the square-angle members are square to both wall surfaces. The nib rule is then fixed in position so that the outside edge of the mould nib bears on the edge of the nib rule.

The screedless method is more suitable in modern plastering as putty and plaster screeds would have to be removed before the finishing coat is applied. This would also apply to moulded angles run in Keene's cement; all gauged lime/plaster would have to be cut away before the moulding was run. A strong plaster/sand core is usually necessary and this must be well keyed. The finishing run will be as for RUNNING IN SITU. A setting member on the running mould will supply screeds for the setting coat on the two walls and, when the running is complete, all rules are removed, the edges cleaned and the bird's beak member formed at top and bottom (Fig. 2). See also BEAD.

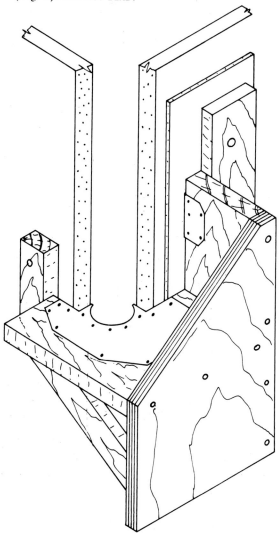

Fig. 2. Angle bead (solid): running

ANGLE, INTERNAL, COVED

See CORNICE; SKIRTING.

ANGLE, INTERNAL FORMED, INTERNAL PLASTERING

This is the natural internal angle formed by two walls meeting. The correct method when FLOATING is to rule the angles out to a straight surface and, before the material hardens, cut out to a clean sharp intersection by passing the laying-on trowel flat down the wall surfaces and allowing one corner of the trowel's toe to cut into the opposite wall surface. All surplus floating material should come away.

When the FINISHING coat is applied, all angles are again ruled out. Just before the finishing plaster hardens fully, a wet crossgrain float is passed up and down the angle and finished with an internal-angle trowel or twitcher.

ANGLE TEMPLATE

See ANGLE, EXTERNAL, FORMED, internal plastering.

ANTIQUE PLASTERWORK

This entry deals with the problems of repairing, restoring and preserving old lime plasterwork, i.e. all moulded plasterwork executed *in situ* before the introduction of modern fibrous plastering techniques, and the sand–lime undercoats and finishes which preceded gypsum and Portland cement. Although plaster of Paris has been known in England since 1254, Portland cement since 1824 and various other types of cement since before that date, the undercoats for plastering were essentially sand–lime mixes containing hair for reinforcement. The finishing coats were also sand–lime mixes which, for very high-class jobs, were sometimes gauged with plaster of Paris. These mixes were still being used well into the present century.

Before undertaking restoration work, the plasterer must have as complete an understanding of the nature of old plastering as he has of modern. He will find himself confronted with a comparatively hard and brittle finishing coat over a crumbling friable mass which is coming away from its background and which cannot readily be taken down without it breaking into small pieces. The background itself will probably be brickwork or wood lathwork, this latter being used not only for ceilings but for timber-stud walls. Occasionally other backgrounds may be encountered, e.g. crude lime concrete reinforced with straw, but these were used in the construction of working-men's cottages rather than stately homes. Lime plastering on wood lathwork was also used to face buildings externally.

Although ornament was modelled and carved *in situ* with lime stucco mixes, small plaster of Paris casts were also used as far back as Elizabethan times. Anything but the smallest mouldings were always run *in situ* with templates, but ceilings consisting of small repeating patterns, such as ornamental panels enclosed by strapwork, could be formed from plaster moulds. An original was modelled in clay or modelled and carved in a conventional plaster mix. A plaster mould was then cast from this model. The mould was sealed and a release agent applied before each cast was made. Plaster of Paris was then poured into the mould and the mould immediately wedged in position so that the back of the cast adhered to the background. The mould was removed when the plaster had set. A whole ceiling could be built up in this way and the joints between the casts subsequently touched up. No moulding with undercut could be reproduced in this way but, if required, could be carved in afterwards.

This advance in the production of moulded work was seen at the time as detrimental to the art of plastering, moulding by a sterile, mechanical process replacing the living character of flowing ornament. In the course of time, moulding techniques developed with the introduction of sulphur moulds and gelatine moulds. Introduced around 1850, gelatine as a moulding material was indeed a great advance. Being pliable, it allowed the reproduction of undercut models without the need for plaster piece moulds, and the reproduction of enrichment – such as leaves and flowers – without the need to carve in the undercut relief afterwards. Robert Adam dressed his plasterwork with GESSO in which his fine detail could be efficiently carved. Through Georgian and Victorian times it became policy to run mouldings containing beds and to cast the lines of enrichment in short lengths and plant them in these beds.

In spite of these developments in finish, there were no corresponding developments in undercoats. A major problem with a sand–lime mix applied to lathwork is that the KEY is the only thing securing the plaster coat to the basic structure. A house is moved frequently during its lifetime (by anything from natural vibrations to bombs); the longer the life the greater the amount of movement, but even late Victorian plastering has been sufficiently disturbed for its key to have been broken. Thus, much old plasterwork is no longer secured to its background. See also REPAIR AND RESTORATION; SQUEEZE.

ARCH, FIBROUS

Plain moulded

Where all or part of an arch can be fixed as a double-sided unit, the mould can be made up and the casts taken from it in the same way as for curved BEAM CASE. The reverse mould required to cast an arch in half and with the joint in the soffit can be spun or run down on the bench as shown, either as a single piece (Fig. 3 (b)) or as a slip-piece if features (such as a sunken-panel soffit) provide undercut. Figure 3(a) shows the arrangement for the slip-piece

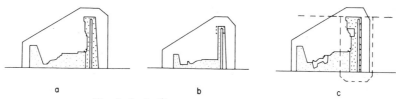

Fig. 3. Arch (fibrous): run reverse moulds

mould. See MOULD, REVERSE, run slip-piece. Either piece can be the loose-piece. Where the riser producing the soffit is loose, it can be reinforced after running by wadding on 25×50 mm battens on the back.

Enriched

Reverse moulds may either be taken from a model of the half-section as run- or clay-case moulds, or have the plaster pieces run containing beds to receive existing or purpose-made insertions. The insertion in the soffit in (c) provides excessive overhang when the riser is run or spun as above. It may, however, be run upside-down as shown and reinforced before turning over. This may be considered preferable to making up a drum, which involves far more operations. Nevertheless, a case mould over a drum is used when the cast cannot be joined down the centre, e.g. one with a fully enriched soffit. See BEAM CASE, to circular or elliptically vaulted ceilings.

Casting, fixing and stopping are as for BEAM CASES.

ARCH, SOLID, INTERNAL

Types

1 Through arches with MOULDINGS run on both faces of the wall and the SOFFIT between.
2 A false arch, usually over a recess in a wall, with a soffit and moulding on one face only.

With the false arch the whole soffit can be run off with, or separate from the face moulding either from a pivot or from the recessed wall.

Methods

1 Running with a GIGSTICK.
2 Running with a PEG MOULD and RIB.
Arch outlines based on circular curves up to a reasonable size may be run with the help of a gigstick. Arches based on elliptical curves or very large circular curves should be run on a rib, using a peg mould.

In all run arches, it is also necessary to run up part of the soffit with the arch running mould. Whether the soffit is plain or moulded, sufficient must be run up to act as screeds when either forming or running the remainder of the soffit. Sunken-panel soffits will have stiles returned across the soffit at the springing line.

Formed arch

Arches of this nature are always plain, both on the sides and the soffit. They can be single- or double-sided and can appear over openings or be a decorative feature superimposed on a wall. A timber rule may be bent round the curve but the most accurate method is to cut two hardboard or plywood TEMPLATES to the curve of the arch. When the floating coat is firm the templates are fixed in position, one either side of the opening. They must be levelled through and, if possible, squared from the wall surfaces. The soffit is then floated out to within 3 mm

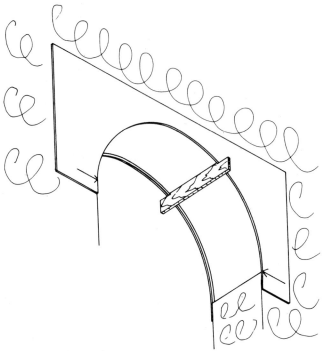

Fig. 4. Arch (solid): ruling in from templates

of the template edges: a gauge may be used. When the floating has set, the finishing coat is applied and the templates may be removed after it has set hard. They may be levered away from the wall slightly so that a small selvedge is formed to act as a SCREED when setting the wall surfaces and assist in forming a good ARRIS.

Gigstick method

A piece of stout timber, not less than 100×50 mm, should be cut so that it will fit inside the arch opening just below the springing line. At the centre of this a pivot block complete with centre pin should be fixed. Two temporary blocks are then nailed to the walls forming the arch returns, positioned so that when the bearer is laid on them the centre pin

will level in accurately with the springing line. The cross-piece is the bearer or stretcher and should always be used 'on edge'.

Before all the fixings are made permanent the running mould, complete with gigstick and FISH-TAIL, should be offered up to the correct position so that adjustments may be made to ensure the running will be both plumb and square.

Once all of this has been done, the formation of the putty and plaster running screeds can take place. This is done by applying two vertical screeds, one either side of the arch opening and continuing above the crown. When these have set, the slipper screed should be RULED OFF the two vertical screeds. All three must be perfectly flat and plumb. The running of the moulding will be in either lime putty and plaster or Keene's cement. Large sections should be CORED OUT and KEYED.

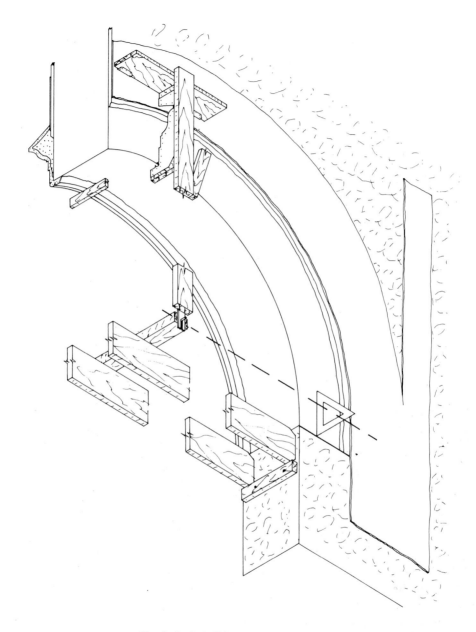

Fig. 5. Arch (solid): spinning a semi-circle

Gigstick method – multi-centred (Fig. 7)

The centres may be mounted on a bearer, on the wall surface or on a large timber plate fixed across the opening. The procedure for the screeding should be as for normal gigstick run arches. At intersections it will be necessary to run part of the initial run on a lime-washed backing so that it may be cut, taken down and placed in a safe position.

When the second length has been run past the line of intersection and cut perfectly to the mitre line, the cut-down section is cut to fit and planted in position. This process eliminates the need for large mitres on curved runs. Another possibility is the running down, on a spot board or bench, of small curved runs. These are then cut accurately and fitted between run *in situ* lengths, again cutting down on the mitring.

As with spinning other curved mouldings where the same section continues through different curves,

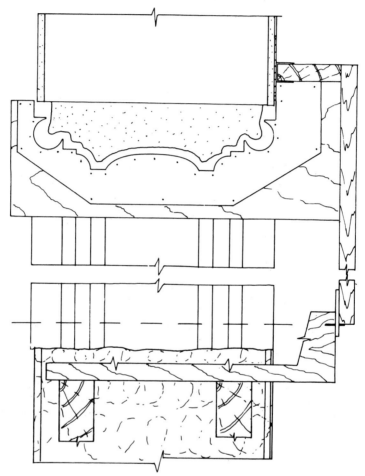

Fig. 6. Arch (solid): spinning a semi-circular moulded soffit

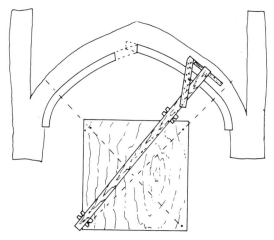

Fig. 7. Arch (solid): spinning a multi-centred Tudor

it is possible to complete this in one application. By fixing pivot blocks of the same height and fishtails so that they pick up from one centre to the next without a jump, a sweep consisting of two or three differing radii can be run in one. The multi-centre Tudor arch is typical of the twin-radii method.

Figure 8 shows the running of an equilateral Gothic arch. The soffit will prevent the face moulding being run through at the crown and cut down. The soffit is therefore run off first, the face moulding being masked off on the running mould. When the soffit portion of the stock is removed, the face moulding can be run through.

Peg-mould method – from inside of arch

This will require a purpose-made rib of either wood or fibrous plaster. The rib must follow the line of the desired curve accurately. However, they should be constructed in such a way that the actual rib curve is smaller than the arch outline by the distance from the outline to the line followed by the pegs. The pegs may be driven into the front edge of the slipper or be blocks fitted to its upper face.

In both cases the peg projection should be sufficient to act as a rabbet. The positioning or spacing of the pegs should be so that the profile cuts as perfect a normal to the curve as is possible. The screeding will be as for the gigstick arch and the running under the material in which the arch is to be run (Fig. 9).

Peg-mould method – from outside of arch

A hanging mould may be used from a running rule fixed on the face of the arch (Fig. 10).

ARCHITRAVE

The lowest member of the classical ENTABLATURE. The moulding surrounding a door, window or any rectangular opening. The picture rail forming the frieze and architrave to a cornice on an internal wall.

ARCHIVOLT

An ARCHITRAVE round a curved opening.

ARRIS

The sharp angle formed by the meeting of two surfaces where an external angle is formed.

ASHLAR

See FINISH/RENDERING, EXTERNAL.

ASTRAGAL

See MOULDING, SECTIONS.

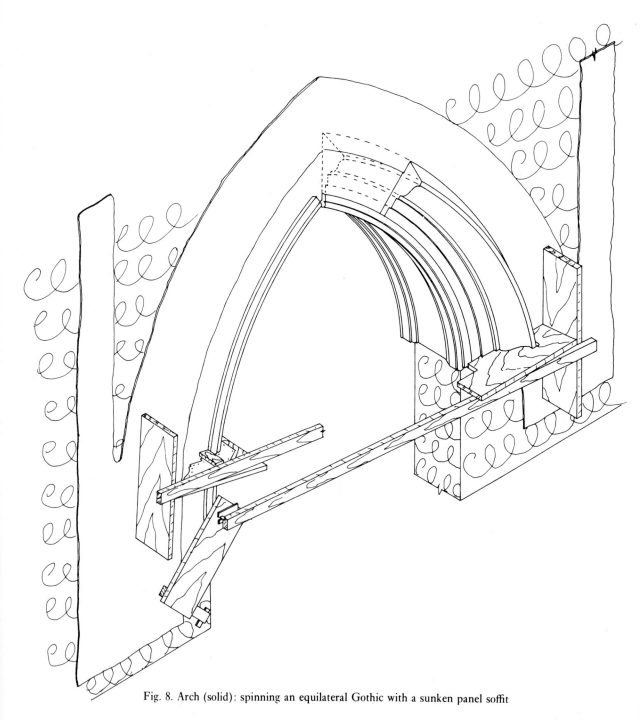

Fig. 8. Arch (solid): spinning an equilateral Gothic with a sunken panel soffit

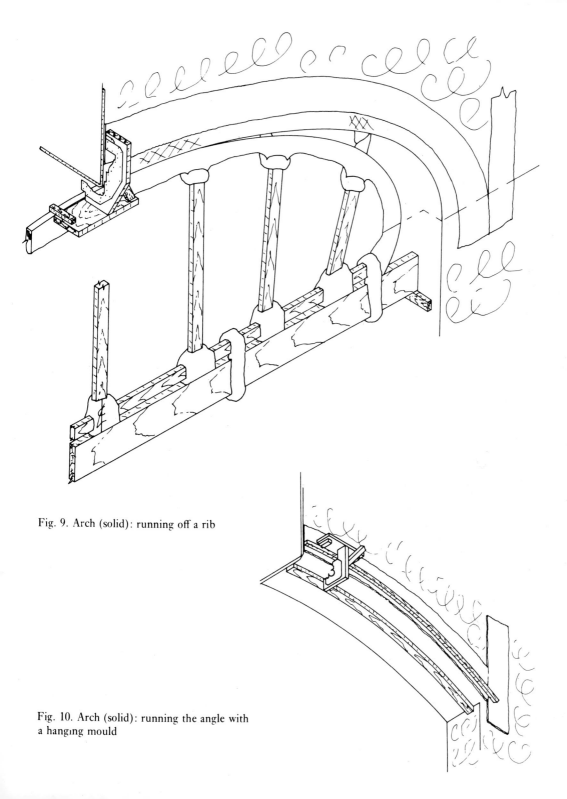

Fig. 9. Arch (solid): running off a rib

Fig. 10. Arch (solid): running the angle with
a hanging mould

13

BACK

A plaster cast over the back of a skin mould, flood mould or clay squeeze. See MOULD, REVERSE; REPAIR AND RESTORATION.

BACK-AND-FRONT MOULD

See under MOULD, REVERSE.

BACKED RULE

See RULE.

BACKGROUND

A background is the surface to which the first or only coat of plaster is to be applied. As far as the plasterer is concerned there are only two types of background, low suction or high suction. Within these two, however, are many different kinds and sometimes a mixture of two or more. In general terms, with gypsum plaster and sand backings a stronger mix is used on the low-suction backgrounds. See also specifications in entries below.

Background preparation

All backgrounds must be free from oil, grease, dust and dirt. Solid-built backgrounds, concrete, brickwork, blocks and the like should have raked joints and contain no loose particles. They should be lightly dampened before the plaster is applied.

Lathings must be checked for correct fixings and contain no broken or damaged portions. Ends of metal lathing sheets and tying wire must be turned away from the surface to be plastered.

Bonding agent

Any background treated with a bonding agent must be recognised as a low-suction background, and BACKINGS specified accordingly. Many of these are unsuitable for external plastering, and those that are must be used in accordance with the manufacturers' instructions. In general they should be mixed with Portland cement and sharp sand and applied as a spatterdash coat when used externally. Internally they must be applied to a clean background with a brush, and plaster applied within the time limit demanded by the manufacturers.

Bricklath

A pliable metal lathing consisting of a square-mesh lath with small pieces of burnt clay at every intersection of the crossed wires. The burnt clay is similar to common brick and will supply a reasonable amount of suction when being rendered. It is made in sheets of varying sizes from about $2\,\text{m} \times 600\,\text{mm}$ and must be fixed across the joists with galvanised nails or staples at 100 mm intervals.

Brickwork

Common clay brickwork is a high-suction background; more dense brick, such as engineering brick, is a low-suction background.

Building blocks, lightweight

Manufactured from waste products combined with an air-entrained cementatious mix. These blocks are suitable for partitions and the inner skin of cavity-wall construction. They are a high-suction background and backing coats specified accordingly. Thermalite blocks are a typical example of this material.

Clinker blocks

Often referred to as breeze blocks, these may be used as partition and inner-skin linings in cavity-wall construction. A low-suction background.

Composite backgrounds

Backgrounds that consist of two or more materials. Cracks may occur on the plaster surface at the joints due to differing expansions. One way to eliminate the possibility of cracking is to isolate the concrete or wooden members by covering them with heavy building paper lapping over the brickwork by a minimum of 25 mm. It is nailed at its edges to the brickwork joints with galvanised nails. The paper is then covered completely by expanded metal lathing fixed in the same way. The lathing should then receive a scratch coat of metal lathing plaster. The remainder of the wall is then floated in browning plaster with the exception of the scratched metal lathing plaster. This portion should be floated in metal lathing plaster before the remainder has set, and all ruled in as one.

Concrete (smooth, no fines) and concrete bricks

All mould oil must be removed from any concrete surface before plastering or applying a bonding agent. When plastering direct to concrete using a light-weight bonding plaster, the concrete should be brushed free of dust and then dampened. This will displace air from the surface pores of the concrete and ensure good adhesion between the background and the bonding plaster. The plaster should be applied in several thin coats at the same time; the recommended thickness for a backing coat of light-weight bonding plaster on concrete is 8–11mm.

Cork

A low-suction background. In modern specification a lightweight bonding plaster is normally used as a backing coat and the finishing grade of a similar plaster used for setting.

Where this finish is not considered sufficiently hard two coats of Class C finish have been used, with a period of 24 hours between coats. For even stronger, harder and more durable construction, the cork is surfaced by tightly stretched galvanised wire, stapled every 100 to 150mm before plastering.

Cork is most frequently used as a thermal insulation material and many large refrigeration units are lined with this material and then plastered. Cork slabs have also been used to line certain types of concrete domestic dwelling.

Expanded polystyrene

Usually bonded to concrete or fixed to timber supports as for plasterboard. This material improves thermal insulation. All joints and angles must be scrimmed with 90 mm hessian, bedded in Class B board finish plaster.

Alternatively, expanded metal lathing or galvanised chicken wire may be stretched tightly across the face of the polystyrene, nailed or stapled to the background or supports and floated in lightweight metal lathing plaster. A low-suction background.

Fibre boards

These are not considered a perfect background for plastering. When fixed to joists or studding the spacings should not exceed 300 mm for the 9 mm boards and 400 mm for the 12.5 mm boards.

All board edges around the perimeter must be supported by additional joists or noggin pieces. The boards must be nailed across each joist at 100 mm intervals with 30 × 2.6 mm galvanised clout nails. A 3 mm gap is necessary at all board edges, and the joints staggered to avoid long uninterrupted joints. They must all be scrimmed as for plasterboards and the specification should be as in finishing plasterboard, one or two coats.

A low-suction background.

Gypsum lathings/plasterboards

(Appendix 3, Table A.3.)

Applications

1. Wall linings

(a) Boards bonded directly to solid walls with bonding compounds: minimum overall thickness 22 mm. See DRY LINING

(b) Boards screwed to vertical metal firring channels fixed to solid walls by bonding compounds or plugs and screws: minimum overall thickness 25 mm.

(c) Boards screwed to vertical metal studs standing in metal channels fixed to floor and ceiling. Metal brackets to the wall are used as intermediate fixings for the studs: overall thickness 40 to 140 mm.

(d) Boards screwed to vertical metal studs that are large enough in section to span between wall and ceiling channels without the need for intermediate bracket support to the wall. Suitable for soundproofing walls.

2. Partitions

Timber studs

(e) Boards screwed to 38 × 75 mm minimum vertical studs spanning between beams fixed to ceiling and floor.

Metal studs, single frame

(f) Boards screwed to a framework comprising vertical metal studs spanning between metal channels fixed to ceiling and floor.

Metal studs, double frame

(g) Two metal frames as described in (f) fixed independently of each other a small distance apart, thus increasing the thickness of the partition and the distance between the boards and giving better sound insulation.

3. Ceilings

Timber joists

(a) Boards screwed directly to timber joists.

(b) Boards screwed to metal sections fixed underneath and at right angles to timber joists.

Acoustic floors/ceilings

Boards screwed to a 'resilient' metal section fixed to the underside and at 90° to the timber joists, impeding sound transmission by reverberation. For the best results, a plasterboard plank, supported by metal channels fixed to either side of the timber joints, fits between and flush with the tops of the joists. The floor decking is held off this structure by a 'resilient' isolation section fixed along the top of the joist.

Suspended ceilings

(a) In a *single grid* system boards are screwed to a metal framework comprising a single layer of parallel metal bearers, each suspended from the background soffit by metal hangers.

(b) In a *double grid* system primary parallel metal bearers are each suspended from the background soffit by metal hangers and carry secondary parallel metal bearers underneath and at right angles to them.

(c) For a *demountable* ceiling grids may be formed using T-bars or metal sections to create a lay-in system.

4. Fire protection for structural steel members

Systems exist, both with and without the use of

framing members, for encasing stanchions, beams, etc.

Fire protection for ducts

Such systems also exist for encasing ventilation ductwork and electrical cable trays.

Fire and acoustic ratings

If a fire-rated system is to be installed it must be built identically to the one built and tested in the fire testing station. Identical materials and components must be used. Therefore it is essential that the manufacturer's specification is obtained and followed strictly. This includes methods of construction together with the grids and support systems, as well as materials: grid components; type of board and number of layers; ancillary components/rock wool/glass-fibre insulation; caulking (mastic etc).

PLASTERBOARD FIXING

(See Appendix 3 Table A4.)

PLASTERBOARD, FIXING SYSTEMS

These comprise various forms of: screws and plugs, metal stud and channel sections, metal hangers and rods, metal clips and brackets, movement-joint sections, metal corner jointing sections, and shadow gap jointing sections. Detailed information of all these may be obtained from the manufacturers of the plasterboards.

Metal lathing, steel for plastering

The following types of metal lathing can be used as a background for solid plastering.

1 Plain expanded.
2 Ribbed expanded:
    (a) with ribs forming an integral part of the expanded sheet,
    (b) with ribs attached subsequent to expansion of the sheet.
3 Perforated.
4 Dovetailed.

The blank sheet from which the lathing is made is of mild steel. The expanded types are made by slitting and expanding a blank sheet so as to form a continuous network of small mesh. The perforated type is made by piercing holes in a blank sheet to form perforations so as to supply a satisfactory key. Dovetailed lathing is made by pressing steel sheets so as to form a series of dovetailed-shaped projections that again will supply a satisfactory key.

Sheet sizes: not less than 2000 mm, not more than 3675 mm. The width is usually 700 mm.

FIXING EML

To timber joists, 38 mm galvanised nails or 32 × 2.032 mm galvanised staples at 100 mm intervals and always across joists, straining the sheet taut both ways.

To steel channels, as above but using 18 g or 1.219 mm galvanised, soft tying wire. Make a hairpin of the wire, push through the mesh, up and over the runner. Pull both ends of the wire together and twist three or four times before cutting the ends off with a pair of top cutters, then push flat.

End laps in both cases wired at 150 mm intervals. The laps must be 50 mm where they occur on bearers and 75 mm where they occur between bearers.

Side laps should not be less than 25 mm and wired at 150 mm intervals.

Average *size* of EML mesh 5 mm SWM, average *gauge* of metal 24 g or 629 mm, average *weight* per m^2 1.63 kg.

FIXING RIBBED LATH

To timber as with EML, nail through the rib.

To steel as with EML but the 18 g wire must be doubled.

End laps 50 mm on supports and 100 mm between supports; here ties should be placed at least two to every rib.

Side laps, overlap ribs and tie every 150 mm. Again always fix across the supports.

See also HY-RIB.

Background

Table 1 British Gypsum systems

Background/lining	Carlite Browning	Carlite Tough Coat	Carlite Bonding Coat	Thistle Hardwall	Thistle Renovating	Thistle Universal One Coat	Thistle Projection	Thistle Multi-Finish	Thistle Board Finish
	TWO COAT					ONE COAT			
	THICKNESS IN mm								
Gyproc Baseboard, Gyproc Lath, Gyproc Wallboard, and Gyproc Plank			8			5	5		2
Common brick walls	11[d]	11		11	11	11	11		
Concrete bricks with raked joints	11	11		11	11	11	11		
Engineering bricks/calcium silicate bricks, both with raked joints		11[a]	11	11[a]	11[a]	11[a]	11[a]		
Dense aggregate concrete blocks		11	11	11	11	11	11		
Lightweight aggregate concrete blocks	11[d]	11		11	11	11	11		
Aerated concrete blocks (pre-treatment may be necessary to control high suction)	11[d]	11		11	11	11	11		
No-fines concrete	11	11	11	11	11	11	11		
Normal ballast concrete		8[b]	8[a]	8[b]		10[a]	10[b]	2[a]	2[a]
Other aggregates concrete		8[b]	8[b]	8[b]		10[b]	10[b]	2[b]	2[b]
Precast concrete units/composite ceilings			8[a]			10[a]	10[b]		

Contd

18

Table 1 *Contd*

Background/lining	Carlite Browning	Carlite Tough Coat	Carlite Bonding Coat	Thistle Hardwall	Thistle Renovating	Thistle Universal One Coat	Thistle Projection	Thistle Multi-Finish	Thistle Board Finish
	TWO COAT					ONE COAT			
	THICKNESS IN mm								
Expanded metal lath	11[c]	11[c]	11[c]			11[c]			
Expanded metal lath spray lath only							11[c]		
Undecorated backgrounds treated with PVA bonding agents in accordance with manufacturer's instructions (eg glazed surfaces)	8		8	8				2	2
Backgrounds treated with proprietary bonding fluids (eg glazed surfaces)			8			10	10	2	2
Expanded polystyrene soffits			8[a]			10[a]	10[a]		
Expanded polystyrene walls			11[a]			11[a]	11[a]		

[a] PVA bonding agents may be required
[b] PVA bonding agents are required
[c] From face of lath
[d] Of moderate suction

Gypsum plasters are available from other manufacturers. For information on Knauf plasters and British Gypsum finishing plasters, please see Appendix 3. Lafarge Plasterboard produces Supreme skim plaster for internal use only on plasterboards. Snowcem produces Snowplast Universal, a regarded hemihydrate one-coat plaster containing perlite and other additives; it is for internal use on most 'suitably prepared' backgrounds, including brickwork, blockwork, concrete and plasterboard.

Nail through the ribs, otherwise as EML.

FIXING DOVETAILED METAL LATHING
As for expanded metal lathing. This is also referred to as corrugated metal sheeting.

K lath

TYPES: AQUA K LATH
Welded wire/fabric lath 50 × 37.5 mm mesh. An absorbent perforated paper is fabricated between back and face wire. Waterproofed building paper is laminated to the back of the absorbent paper. 50 mm flaps are provided at one end and one side of each sheet. May be used internally or externally for waterproof renderings.

GUN LATH
As for Aqua K Lath, but without the waterproofed building paper and flaps. Used on interior walls, ceilings and partitions.

FIXING
Long dimension across studs or supports, bend around corners and pick up to first support. Horizontal joints lapped 50 mm, vertical joints – minimum 37.5 mm, all vertical laps staggered. Fix as for metal lathing, including tying. Suitable for hand or spray plastering.

Newtonite lathing (damp-wall treatment)

A pitch-impregnated, fibre-based corrugated sheet lathing. Available in rolls 1000 mm wide × 5000 mm long.

Newtonite must be fixed to clean brickwork or masonry with the corrugations vertical and the metal strips showing on the side to be plastered. Three rows of nails in the raised corrugations are necessary, one row near to each edge, top and bottom and one row in the centre of the sheet, spacings 200 mm, nails galvanised clout or masonry. A small gap should be left at the top to allow air circulation. Alternatively air bricks can be let into the external wall.

Laps of 100 to 150 mm are necessary at all vertical joints and the horizontal joint should be butted with a 100 mm strip of bitumen felt fixed behind it.

Where the floor is solid the lathing should be raised about 50 mm from the floor and either a waterproofed floor topping lapped over the edges, or a polythene membrane fixed to the wall surface at least 150 mm above and behind the lowest edge of the lathing.

This is a low-suction background and a lightweight aggregate metal lathing plaster is recommended for the scratch and floating coats. Alternatively a Class B browning plaster-sand mix may be used for the backing coats.

Tiles, clay (pots)

Hollow clay tiles with grooved surface, frequently used as ceiling backgrounds in concrete-constructed buildings. A high-suction background.

Tiles, glazed

Each tile must be checked for soundness, and good adhesion must exist between the tile and its backing. All grease and dirt must be removed and tiles should be treated with a bonding agent before being plastered in lightweight bonding.

Alternatively they can be spatterdashed for cement–sand plastering. A low-suction background.

Synthaprufe

Often used as a background for waterproofing, a black substance with the consistency of tar.

It is painted on to the sub-background and dashed with dry sharp sand. This forms a key for the backing coat. The same process may also be used for plastering over old glazed tiling; the tiling must be sound.

Undercoat plasters

Thistle Dri-Coat is a premixed, lightweight cement-based plaster, incorporating expanded perlite aggregate. It is intended for internal use after the installation of a new damp-proof course or after remedial treatment in cases of penetrating damp. It contains special additives that resist the passage of efflorescent and hygroscopic salts. When it has dried it forms a barrier to the passage of moisture but not to water vapour, so that any remaining dampness in the building can dry out.

It should not be used below ground level because of hydrostatic pressure. Backgrounds are suitable DPC-treated substances such as brick or stonework and the coat thickness is 11 mm. The finishing coat is either Thistle Board Finish or Thistle Multi-Finish. See Table A.1 in Appendix 3.

Wood laths

For many years the traditional background for plaster ceilings and partitions. Today the lath is used mainly for reinforcing fibrous plaster casts. Originally the laths were split from specially selected blocks of straight-grained hardwood. This was one reason for the hatchet blade on the lath hammer, and this type of lath was known as a riven lath.

The sawn lath that we know today is available in three *grades*: lath 3 mm, lath and half 6 mm, double lath, 9 to 10 mm thick. All laths are 25 mm wide and the lengths can vary from 1000 to 3000 mm; longer ones are available if required.

When used as a background for plastering, the lath and half should be used. It must be fixed by a method known as the butt and break joint. In this, the laths are nailed across the joists with 25 mm lath nails, the gap between laths being 8 mm. The ends of the lath are butted and the line of the end joints is staggered by moving it to the next joist every 1000 mm. A broken joint should eliminate the possibility of long straight cracks appearing.

Wood-wool slabs

Made from compressed wood-wool fibre and Portland cement. Used either as permanent shuttering to concrete ceilings or may be fixed to wooden studs to form a building lining. Have excellent sound-resistant qualities.

All joints and angles must be reinforced either by hessian scrim bedded in Class B board finish, or metal scrim applied by a dispenser.

A low-suction background providing a good key. See also PLASTER, GYPSUM.

BACKING

Undercoat.

BALUSTER

A small column- or pilaster-like support, generally urn-shaped with the basic outline of its body a concave neck swelling into a ball below.

Model

The round section is turned up in a TURNING box (Fig. 124). The square blocks top and bottom are cut and rubbed up as DENTIL BLOCKS from a band either run or cast between two rules.

Mould

When assembled, the two-piece mould is made with the joint across a diagonal of the blocks (Fig. 11). It may be of plaster or glass fibre. Longitudinal reinforcement, preferably angle iron, is positioned to the take clamps when the particular method of casting requires the two halves of the mould to be clamped together. Two-piece clay-case moulds can also be used and these are essential when the baluster is enriched. The disadvantage of these is the poor pick-up of members between the two halves.

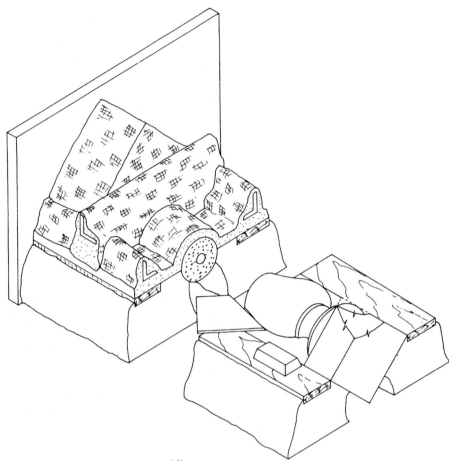

Fig. 11. Baluster (fibrous): plaster piece-moulding

Casting

In either plaster, glass fibre, or cement mixes as for PIECE MOULDS. For a multi-sided baluster see MITRE, RUNNING UP.

BANKER

A platform or base on which plastering materials are mixed on site, and that can be kept clean and free from soil, etc.

BARREL

See VAULTING.

BARYTITE X-RAY PLASTER

See X-RAY PLASTER.

BASE

The lower part of a classical column or pilaster on which the shaft rests.

BASEBOARD

See BACKGROUND.

BAUXITE

See CEMENT, HIGH-ALUMINA.

BAY

An area formed by splitting up a larger area with features – beams, piers, mouldings, etc. – therefore generally appearing as a recess.

BEAD AND REEL

See under ENRICHMENT.

BEAD, METAL, EXTERNAL PLASTERING

Render stop bead

Consists of metal lathing and a splayed strip of galvanised steel sheet. It is used to achieve a neat edging to cement renderings as in BELL CAST. The method of fixing is by galvanised or masonry nails into the background, passing through the lathing. This will be covered by the rendering.

BEAD, METAL, INTERNAL PLASTERING

Angle bead

Providing a true arris that should not chip or crack with the modern softer pre-mixed plasters, a hollow metal bead that will form the finished angle to the plastering. It is fixed to the background and worked to as a screed. A thick-bed grade with EML on either side is used for floating coats and a thin-bed grade with perforated metal on either side for finishing coats 3 mm or 6 mm.

The bead should be of galvanised steel or epoxy-coated stainless steel. All damaged and cut edges should be coated with galvanised paint or an appropriate lacquer.

Metal angle beads are usually fixed by applying plaster dabs at 600 mm intervals to both sides of the angle. The bead is then cut, pressed into the dabs and squared in and plumbed. When using thin-bed grade on plasterboard, it is fixed by nailing. The nose or rim of the angle bead will determine the thickness of the backing coat. When this has set, it should be cut back just below the level of the rim so that the finishing coat will be just a little proud of the metal.

On long runs, lengths may be joined using a dowel cut from a length of 8 g (4.064 mm) galvanised wire, or a suitable galvanised nail with its head cut off. This should be inserted into both ends of the hollow bead, ensuring continuity of line.

Corner mesh

Expanded metal lathing in 100 mm strips bent at right angles to assist in the reinforcing of internal angles. It should be embedded in the plaster applied to the two walls that meet to form the internal angle. Another type of corner mesh has two ribs lengthways at the centre of the strip. This is put up before plastering (as with external metal angle bead) and the two ribs are used to form a screed for ruling off to and will also form the internal angle.

Plasterboard edging bead

Available in two sizes: 10 or 13 mm. This is a dual-purpose bead. It can be fitted so that the perforated flange is behind the plasterboard, mainly for dry lining; or with the perforated flange exposed over the face of the plasterboard, in which case it should be covered with board finish or jointing filler. In both cases it should be fixed by nails, and its prime function is the protection of plasterboard edges.

Screed bead

A central galvanised-steel channel flanked by two metal lath wings. It may be used as a screed or as a method of forming a joist between different types of plaster or between flooring and wall plastering.

The normal procedure for fixing is by dabs of plaster. These should not be left behind cement renderings or granolithic work. Lengths may be joined as for angle bead (above).

Stop bead/casing

Metal lathing and a returned strip of galvanised steel. Provides a neat finish to edges of internal plastering. At doorways, it will eliminate the necessity for cover mouldings and is also used between plaster and skirtings and different plaster materials. Fixing may be by plaster dabs or galvanised nails driven through the holes provided. It may also be wired to metal lathing.

BEAM CASE – FIBROUS

A cast, comprising two cheeks and a soffit, to cover a beam or to form a false beam. Although the reverse mould may be cast as a piece mould from a model, comprising two cheeks located in a saddle reproducing the soffit, it is usually made up of run sections. Using this form of mould, the beam case may be built up in one of three ways.

A beam case containing a cornice may be cast in one, with plain sides, and have the cornice fixed to it during the fixing operation (Fig. 12). This method

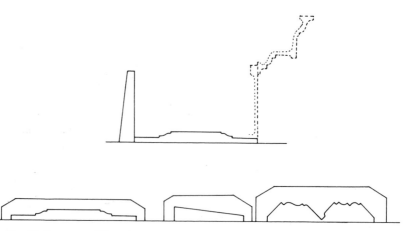

Fig. 12. Beam case (fibrous): the run reverse moulds to cast a cornice separately

is chosen when a section of the cornice that is being fixed elsewhere can be utilised, and obviates production of the full cheeks. The disadvantage is the multiplicity of operations.

cheek, this method is sometimes chosen when only a few cases are required. It is also chosen when the cheek and cornice are completely enriched and so need to be reproduced by a run-case mould. The

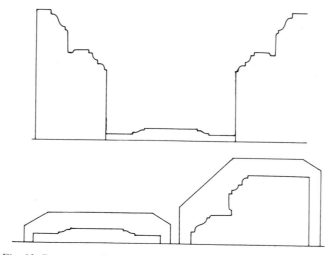

Fig. 13. Beam case (fibrous): the run reverse mould to cast in one piece

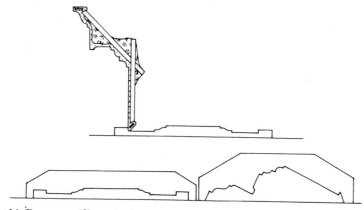

Fig. 14. Beam case (fibrous): the run reverse moulds to stitch the sides with the soffit

Figure 13 shows a mould to produce the whole casement in one, the cheeks or risers being produced as RUN CASTS.

In Fig. 14, the casts of the two cheeks are stitched to the soffit. Because the mould needs only one

disadvantages are that three casting operations are necessary, the casement is much heavier, and making good is always needed between the soffit and each cheek.

Stiles across a sunken-panel soffit are run, fixed

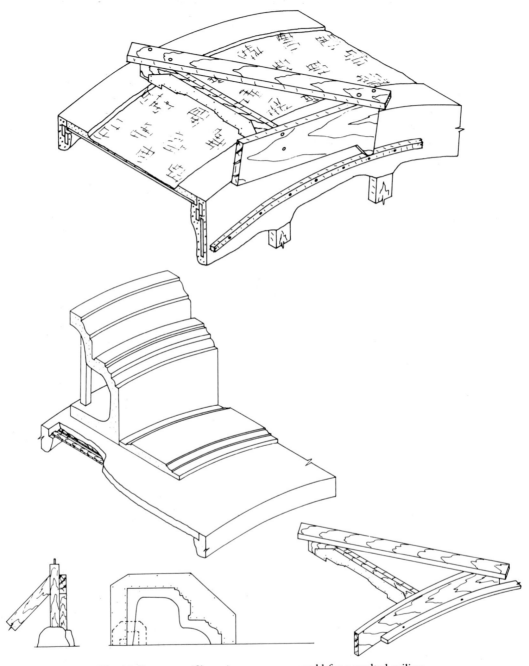

Fig. 15. Beam case (fibrous): run reverse mould for a vaulted ceiling

and stopped in recesses cut in the reverse mould to receive them.

To circular or elliptically vaulted ceilings

Although the methods of Figs 12 and 14 may be employed, the reverse mould is usually made up as in Fig. 15 and assembled over a DRUM. The risers for the reverse mould are spun or run in their projection on the bench. The reverse mould for the soffits is run over a drum made to a curve smaller than the ceiling by the whole depth of the beam-case mould. A circular drum will require either a curved slipper or a peg mould, whereas, if the stock is to be kept at a normal to a changing curve, an eccentric rule will have to be used, fixed to the side of the drum on a RIB cast specially deep to receive it (Fig. 15).

Setting out

The ceiling line of the beam is set out to the curve of the ceiling. With beams to changing curves, the

The bottom of the riser will be the inside edge of the rule.

Enriched

Lines of enrichment in either the cornice or soffit are probably best provided for as run insertions (Fig. 16), especially when the reverse mould for the rest of the cornice can be designed with the beam case in mind. In this case the metal plates that ran the beds for the insertions are fixed in position in the sections that will run the beam-case mould. Where the cheeks are fully enriched they may both be produced as run cases from a model run and dressed separately on the bench. See MOULD, REVERSE, insertion – run case.

Casting

Casting is carried out as for CORNICES. Where there is sufficient room on fixing, all the brackets, in both cheeks and soffit, should be on edge. When the cast is being fixed by nails or screws to timber battens

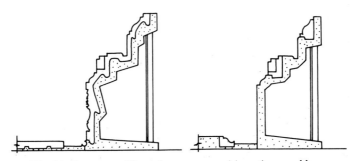

Fig. 16. Beam case (fibrous): run case and insertion moulds

best way of ensuring that the drum fits the risers is to use the following procedure:
1 With the chosen set-up, run the risers.
2 Superimpose on the stock the profile to run the plaster rule off which the ribs are cast.
3 Run off the plaster rule, using the same set-up.

fixed to the beam, the fixing laths will need to be struck off with a gauge to an accurate thickness. If the width of the casement so requires, fixings are also made along the soffit. Extra cross-brackets are incorporated as shown, in order to reinforce the casement's sides during storage and transport. They will be cut out immediately before fixing when the casement is to fit round a beam, but left in when the beam is false. (See Fig. 17.)

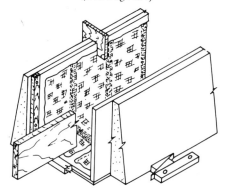

Fig. 17. Beam case (fibrous): forming the strike-offs

Fixing

Figure 18 shows the basic method of fixing to timber or metal. The procedure is the same as for fixing CORNICE.

BEAM, SOLID

See FLOATING TO BEAMS.

BEARER OR STRETCHER

A wooden member fixed in position by the plasterer to assist with the formation of run work in solid plastering, usually where a pivot block is required. See also SPINNING.

BED

A recess in a moulding in which ENRICHMENT is to be planted.

BEDDING

The fixing of plaster moulding or mouldings to a background by ADHESIVES, plaster, PVA, etc.

BELL

The largest part of a classical Corinthian and Composite capital, bell-shaped and carrying the main enrichment between the ABACUS and the ASTRAGAL.

BELL CAST OR BELL MOUTH

Where the bottom of an external finish is lipped to form a drip at plinths, skirtings and other openings so that water does not run down the surface underneath. See also BEAD, METAL, EXTERNAL PLASTERING.

BENCH

Column bench

A column bench has the form of a half-column or column turning box on legs in which column moulds are turned. For construction see COLUMN.

Glass-topped bench

Such a bench has a sheet of glass as its surface and is intended to be used for casting only. It will produce a PLAINFACE with a perfectly flat surface, but the mirror finish is removed when it is cleaned up with a joint rule and busk, as is necessary with all plainface casts in order to remove the grease from the surface–if nothing else.

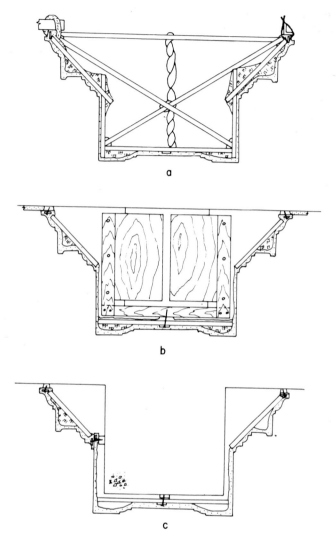

Fig. 18. Beam case (fibrous): (a) fixing a false beam to a metal grid, (b) fixing to timber cradling round a steel beam, (c) fixing to grounds round a concrete beam

Makeshift bench

For casting plainface or constructing simple reverse moulds consisting of flat surfaces to produce casts such as piers, beam cases, etc., sheets of resin-bonded blockboard or plywood may be used, while planks and boards may be used for running with rebated running moulds.

Octagonal bench

Intended to allow the easy SPINNING of circular

mouldings, DOMES, etc; constructed as plaster bench.

Plaster bench

Plaster is the ideal all-round material for the surface of a plasterer's bench. It has no appreciable moisture movement; it will receive nails; it may be easily cut and made good to, thus allowing it to be incorporated as an integral part of a reverse mould; and it may be repaired and re-topped when necessary.

Size

The length of a bench is determined by the length of the fibrous plaster casts to be produced upon it. Conditions such as storage, transportation and site access will dictate the length of the cast, although in the case of cornice required in varying lengths, a 3 m and a 2 m length may be produced as one by cutting a 5 m cast. The longest cast the writer has dealt with was a 5.300 m beam case. Within the range of 1 to 2 m any width of bench may best suit certain work. In most plastering shops (whatever the most common width) an extra-wide bench of some 2 m plus is to be found. It is a great advantage to have all the benches the same height, i.e. their tops level one with another no matter how uneven the floor, as the gangways between them may be bridged over and plaster-topped to provide a bench of any required width. This is often invaluable for large-radius work, domes or extra-wide sections.

Whereas with a low bench maximum pressure can be brought to bear on a running mould, a high one reduces the strain on the plasterer's back during dressing models, etc. A good compromise height is around 0.850 m.

Construction

The timber frame construction comprises cross-members or joists supported by a main bearer, running down either side, that is in turn supported by the legs.

On this framework two types of bench can be constructed: a heavy-duty bench is produced by close boarding laid over the cross-members and covered with a layer of lightweight building blocks. (Plaster applied directly on to the boards would cause them to twist and warp, and this may happen at any stage during the lifetime of the bench.) Blocks are a good material to back up a plaster top as, being of much the same consistency, they will behave in much the same way. Any gaps between the blocks are filled with thick plaster.

A lighter bench is constructed by spacing the cross-members closer together, at 0.300 m centres, and lathing out with wood lath.

As no elaborate carpentry joints are used, it is advisable to tie the bench together with wires and metal plates, to hold against the expansion of the plaster top and the re-topping at later dates. The sides are tourniqued together with galvanised tie wire, and metal corner pieces are used to tie the side pieces to the ends. The plaster top is brought up in two layers. The side and end pieces are levelled round and the cross-pieces fixed at such a height that, when the boards and blocks are positioned on them, some 30 to 50 mm is left for the plaster top. The bench is ruled in with a wooden floating rule. The first layer is kept some 10 mm below the finish by fixing a rebate on the rule. When this has been ruled in it is well keyed every 30 mm by diagonal undercut grooves cut with a chisel.

The topping plaster should be well rubbed in, as in all such cases, for after a matter of seconds the undercoat will have sucked the water from the plaster, making it too stiff to be forced into the key. Two men, one on each side of the bench, should cut off the plaster when it is in exactly the right stage of its set, by drawing the floating rule smoothly along the side rules. Such a finish, produced by a wooden rule, will be open and woolly. When set, it is filled and made smooth by working creamy plaster into the surface with a 300 to 450 mm joint rule.

The topping slab should be of one gauge of plaster and without hollows filled with subsequent gauges to a feather edge. The edges of such patches will always flake under wear. It is better to cut out and key such areas so that the making good is to vertical sides. Water is never splashed on to obtain a good finish. If a plaster surface is shellacked when too wet, the SHELLAC tends to roll off during the process. Such a surface may have water drawn from it by a sprinkling of dry plaster which is scraped off a few minutes later. The bench can then be shellacked.

Maintenance

When making good, all grease should be removed from the indentations. Damping down the bare pieces of plaster will make a good finish easier to obtain but may weaken the adhesion of the existing shellac. The making good is effected by scraping creamy plaster of Paris over the surface with a 300 mm joint rule. Care should be taken that only the damage is filled and that a thin film of plaster is not left all over the bench. Shellacking should be kept to a minimum, only the repaired areas being shellacked, rather than three coats being applied to the whole surface. Too thick a build-up of shellac will crack and flake.

Re-topping

When the surface has reached such a state that making good will not produce an adequately sound finish, the top 10 mm layer of plaster may be chipped off and the bench re-topped.

BILL OF QUANTITIES

See PRODUCTION MANAGEMENT.

BINDER

This term originally included reinforcement such as hair but is now applied only to the setting or hardening material in a plaster mix.

BIRD'S BEAK

See MOULDING, SECTIONS.

BIRD'S BEAK STOP

See STOP, bird's beak.

BLEEDING

See FAULTS; GELATINE.

BLISTERS AND BLUBBING/BLEBS

See FAULTS.

BLOCK

Wood or metal fixed to a surface to give footing to a brace, folding wedges, struts, etc. See also DENTIL BLOCK; MODILLION BLOCK.

BLOWING

See FAULTS.

BLUBBING

See FAULTS.

BOARD, FIBRE

See BACKGROUND.

BOND

The tie between surfaces due to ADHESION or mechanical KEY.

BOND FAILURE

See FAULTS.

BONDED FLOOR

See FLOOR LAYING AND ROOF SCREEDING.

BONE

To adjudge by eye the suitability of a line, straight or curved.

BOSS

Enriched knob covering the intersection of mouldings. Found mainly on ceilings with RIBS and STRAPWORK.

BOXED LATH

See PLAINFACE, reinforcement.

BRACKETING

In solid work bracketing usually consists of a number of wood or metal members applied across the angle formed by the meeting of the ceiling and wall to which wood or metal lathing or plasterboard can be fixed. This is to save material when forming cornices in solid.

The original method used to do this was to cut wood lath to the required length, and bed in plaster gauged coarse stuff, on both the ceiling and wall, well behind the finished moulding line. This would then be rendered in the same material, well keyed, allowed to set, then the cornice would be cored out

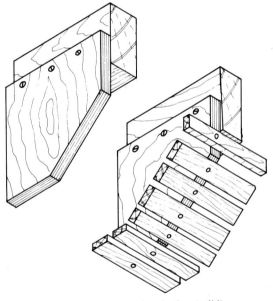

Fig. 19. Brackets and bracketing (solid):
to cornice (internal) – timber

to a muffle. The name given to this method was Scotch bracketing. See CORNICE, Fig. 36.

As wood laths are rarely used in solid plastering now, a more up-to-date method is to fix long narrow strips of plaster board in a similar position with Class B finishing plaster and scrim, then cover the whole plasterboard area with the same type of material. Lime should never be allowed to contact the boards as it will damage the paper.

Framed bracketing is another type used in a similar way. Here the cornice is probably much larger and timber pieces are fixed to the sides of the ceiling joists and to the wall. Wood, metal lath or plasterboard is then nailed to these and cored out in the way described earlier (Fig. 19). For external cornices to be run in Portland cement and sand, metal reinforcing rod is bedded into the wall face below the nib line and above the lowest member line, and expanded metal lathing is wired to these.

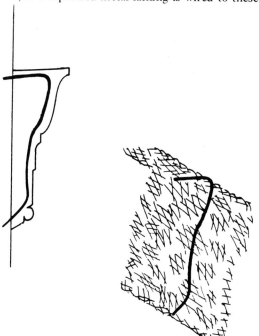

Fig. 20. Brackets and bracketing (solid): to cornice (external) – metal

The spacings should be about 500 mm and the wire tied to the bracket (Fig. 20) with 18 g galvanised tying wire at 150 mm intervals. The wire is rendered with a 1:2 cement:sand mix containing hair or some similar fibrous material to assist with the binding. This is then well keyed and the cornice cored out to a muffle with a similar mix, minus the hair.

The same term can be applied to the cradling of beams and in fibrous plastering a bracket is the term given to the laths laid across a fibrous plaster cornice cast. They can be bruised and laid in to follow roughly the shape of the back of the cast, or wadded in at right angles on large casts behind the wall and ceiling line. A gauge is often used in this case to ensure that they will not project in front of the finished lines. These are termed cross-brackets.

BRACKETS

Cross-brackets

Wood lath reinforcement positioned across the width of a CAST.

Ornamental brackets – truss, console, modillion block

Small blocks may be cut from a run moulding as DENTIL BLOCK. Larger blocks may be ruled in between integral sides which, if moulded, may be formed by joining and mitring spun and run mouldings or, if plain, be of shaped plainface. See MOULD, REVERSE (Fig. 70). The blocks may be dressed by
1 Planting and stopping a capping moulding.
2 Running a moulding over it with a thumb mould.
3 Modelling on enrichment.
The moulds required for these brackets are a piece mould for plain-moulded blocks and a clay-case mould for enriched blocks.

Ornamental brackets, diminished – keystone and truss

The basic block may be formed tapered by either of the above methods, any moulding over it being run with a triple-hinge mould. A flat-faced keystone may be run on the bench as a diminished moulding. The top and bottom are cut at the correct angle to the face, and the variation in depth between the top and bottom is produced by adding an inclined moulding ground.

BRANDERING

This is a very old term, originally applied to the operation of nailing cross-battens to the undersides of joists and at right angles to them when the spacing between the joists was considered too wide to give sufficient support to wood lath ceilings. The battens were fixed at spacings of 12 to 14 inches. Today a similar operation is carried out when the joist spacings exceed the maximum for plasterboards, 400 mm for the 9.5 mm board and 450 mm for the 12.7 mm board. See also BACKGROUND, lathings.

BREAK JOINT

See RIB, casting.

BREAKS AND RETURNS

A break is an interruption in the continuity of the line of a moulding or of plain surfacing.

Internal cornices

When returns are not large enough to be run in position because of insufficient wall line, the procedure is to fix a piece of wide timber to the spot board, at right angles to it, representing the ceiling, and run a short length of cornice on a CORE of wet sand or coarse stuff. This should be run in neat plaster and, when it has set, the small pieces for bedding up in the breaks can be cut from it. They should be well KEYED and bedded in position with neat plaster. The usual minimum size for this is about 100 mm; smaller breaks can be formed freehand, using small tools and joint rules and a spare metal profile to check the shape.

External cornices

When these occur in cement and sand cornices externally they must be worked in by hand. To do this it is necessary to start from the top and work down, plumbing every vertical member in and using a spare metal profile to check the cornice shape. Wooden purpose-made joint rules are the best for this job, and for the finish very small wooden floats are passed lightly over the entire cornice surface. This operation should be carried out at the same time as the running so that the finish is all one.

BRICKLATH

See BACKGROUND.

BRICKWORK

See BACKGROUND.

BROWNING

See PLASTER.

BRUISING

A method used to part-fracture wood laths required to reinforce curved fibrous plaster casts. The soaked laths are hit at regular intervals of about 12 mm with the head of a lath hammer, with sufficient force to bruise the lath but not to sever it.

BUILDING BLOCK

See BACKGROUND.

BULKING

See SAND.

BULLNOSE

See ANGLE, EXTERNAL FORMED.

BURIED RULE

See RULE, ECCENTRIC.

BURNT SHELLAC

See SHELLAC.

BUSK

A flexible piece of spring steel, usually in lengths from 150 to 300 mm and widths from 25 to 75 mm, and sharpened at right angles. It is used to produce a finish on fibrous plaster surfaces by scraping, either dry or lubricated with water.

BUTT AND BREAK JOINT

See BACKGROUND, lathings.

BUTTERCOAT

See FINISH/RENDERING, EXTERNAL, pebble dashing.

CALCINATION

Extreme heating; as a term, can therefore be applied in the course of manufacture of CEMENTS, LIMES and PLASTERS.

CALCIUM SULPHATE

$CaSO_4$. The main constituent of GYPSUM, $CaSO_4 . 2H_2O$.

CAMBER

The concave shape of granite and cement floors and paths. It can also be seen on the outline of a flat ARCH or lintel and occasionally on the SOFFITS to BEAMS. A slight camber or rise will counteract the appearance of sagging. It is similar to the ENTASIS formed on a classical column.

CANVAS

See HESSIAN.

CAPITAL

The crowning feature or ornamental head of a classical column or pilaster. For manufacture see COLUMN; PILASTER.

CAPPING

A cap-like feature formed by a moulding at the top of a pier, etc.

CARBORUNDUM

A compound of carbon and silica. In plastering, carborundum dust is sprinkled on the surface of granite floors before the final trowel. Its main function is to provide a non-slip surface but it also affects the appearance. It must not be trowelled in too heavily and should always be visible on the surface. Up to 1.500 kg/m² may be used. See also FLOOR LAYING AND ROOF SCREEDING.

CARTON PIERRE

A material used in the past to produce fine enrichment. The ingredients were paper pulp, whiting and size. It was mixed to a dough-like consistency and pressed into plaster moulds.

CARTOUCHE

A moulded scroll, frequently containing an inscription, figures or an emblem. Traditionally this was often used to decorate theatre-box fronts.

CASE

A plaster cast supporting a layer of pliable moulding material, together forming a case mould.

CASE HARDENING

See FAULTS.

CASE MOULD

See under MOULD, REVERSE.

CAST, FIBROUS PLASTER

A pre-cast unit consisting usually of plaster of Paris reinforced with wood laths or battens and layers of 6 mm mesh jute hessian.

The timber members are so positioned as to act as struts and braces (in the engineering sense), giving rigidity by imparting compressive and tensile strength, and also providing a means of fixing the cast to its background. In special cases it may, of course, contain galvanised metal such as angle or channel sections, as well as or instead of timber, while the plaster may be Keene's cement or special hard Class A casting plasters. The surface or object from which the cast is made is termed the mould.

In general the term 'cast' is used for fibrous plaster that is to be fixed as a unit of plasterwork. However, we must consider that other objects produced by casting are named according to their function, e.g.

A plaster mould may be cast from a model but is known as a mould.

A model may be cast from a mould but is known as a model.

A case may be cast from a core but is known as a case.

CAST, PLASTER –
CASTING IN FIBROUS PLASTER

Plaster casts may be difficult to remove from moulds for two reasons: (a) binding between surfaces or members; (b) suction/vacuum between cast and mould.

Binding between surfaces or members

This is often encountered with linear run sections. To remove a cast the strike-off down one side is lifted gradually at regular intervals with a hook inserted under the lath reinforcement. A piece of 3 mm lath with a wedge-shaped leading edge is forced into the increasing gap along the entire length. A one-off cast of a slightly undercut section can be strained from the mould in this way, bearing in mind that some fracturing will occur. When, due to design, it is not possible to insert a lath between cast and mould, wedges may be inserted between the lifting-out sticks and the strike-offs on the mould and serve to prevent the cast from slipping back.

The cast is raised by the hook under the timber reinforcement and/or by levering from a block placed on the strike-off of the mould by a file with its point driven under the cast's wood lath reinforcement or by a crowbar inserted under the cast's lift-

ing-out sticks. Removal from cast moulds may also be effected by lifting the whole by the lifting-out sticks of the cast, placing a block of wood on the mould's strike-off to protect it and spread the load, and tapping this block with a hammer.

Suction/vacuum between cast and mould

Suction occurs with certain sections, especially full domes. If the procedures described above should fail, the vacuum is broken by drilling a hole in the top of the cast and forcing water between cast and mould. This can be done adequately by connecting and filling a funnel.

For the removal of plaster casts from wax moulds see WAX, MOULDING; see also COLUMN.

CASTING IN CEMENT

The chief drawback of Portland cement for casting is its slow setting. However, quick-setting varieties may be used.

Wet method

This is comparable with casting in concrete and produces a smooth, dense face on the cast. Moulds are treated as follows – all *porous* moulds (plaster, etc.): shellacked and oiled; *non-porous* moulds (glass fibre, etc.): oiled (this is not strictly necessary with PVC but it prolongs the life of the mould).

Many oils can be used but the best choice is mould oil specially formulated for cement casting. A grout of water and cement is brushed all over the surface, not as a thickness layer as with FIRSTINGS but merely to ensure a perfect fill between the grains of the aggregate. The main mix is then compacted within the mould. Metal reinforcement may be included according to the cast and may be rod, channel, exterior-grade metal lath or weld mesh.

Protruding fishtails may be arranged as provision for fixing. An ideal specification would require all metal to be bronze. The cast is removed when sufficiently strong: this will depend on the type of cement and the section of the cast. Excessive undercut, such as that encountered on Corinthian caps, will make it necessary to leave the cast in longer; this time may be reduced by cutting the PVC, preferably with a joggle knife.

Dry method or pressed cement

With this technique the mix is of the consistency of damp sand which, if compacted into a mould, can be turned out immediately in the same way as children's sand castles. It is obvious that large casts requiring turning over and those with overhang that will collapse under its own weight are unsuitable for this method. The technique produces a texture similar to that of Portland stone. The mix is placed into the mould in layers from 30 to 60 mm thick and beaten down with the ends of pieces of wood shaped to get into all features until it is extremely hard. The result is a dense, close texture on the surface, that is not at all open or sandy.

The texture of the aggregate is of great importance: sharp, angular granules cling together better than smooth ones. The moisture content is also critical: too little and the mass will crumble, too great and it will stick to the mould, causing the face of the cast to be pulled away. Normal release agents will also have this effect and the mould should therefore be made to repel the mix. This can be achieved by either keeping the shellacked mould dry and dusting it with a layer of french chalk before each cast or making it water-repellent by soaking it with paraffin or waterproofing liquid and applying further brushings to the face before each cast.

Moulds must not flex at all as this will crack the cast. This necessitates rigid, one-piece or tightly fitting piece moulds – well clamped together if necessary. Moulds open at both ends (such as squat

BALUSTERS) may be filled in from the top and the pieces withdrawn, the mould being split horizontally into two-piece rings, each ring assembled on the one underneath and filled in in turn. Moulds that are open at the top only (such as piece moulds of BRACKETS) will have to be turned over and the mould removed. To prevent the cast from settling and cracking, it is tamped down slightly full of the strike-offs and a slate or stiff board, on which the cast will remain, is pressed on to the back. While this is held firmly against the back, the whole is turned over. Provision for fixing on site can be made by cutting out a dovetail hole from the back and filling it with damp sand before turning over.

Curing

After 24 hours the dry pressed cast will be at the same stage as a newly struck wet cast. Each should then be repeatedly sprayed with water for 24 hours or until hard enough to be immersed in water, in which they should be left soaking for at least a week.

CASTING IN FIBROUS PLASTER

General method

The main rules that apply to casting are:
1 It should contain a minimum of two layers of canvas.
2 The timber or metal members should be spaced not more than 300 mm apart.
3 All laths should be covered with canvas from above and below.
4 In general, the timber should be placed on edge when the cast is to be fixed by wire and wads to metal fixings, and placed flat when it is to be nailed or screwed to timber fixings.

For casting operations extra to this general method, see under specific item to be cast.

Preparing to cast

All canvas and timber, etc., must be cut to size before beginning casting. The two layers of CANVAS are so arranged that the first covers and overlaps the mould by at least 100 mm in order that it may be 'turned in', i.e. folded back on to the cast, covering the laths round the perimeter. Joints in pieces of canvas are overlapped by at least 30 mm.

All laths must be soaked in water in order that they may fully expand before they are held by the set plaster. If dry laths are put into a cast near its face they will expand and crack it along the line of the lath. The position of every lath will be revealed on the face of the finished cast. In general, laths are cut to size some 20 mm shorter than the space they are to occupy to allow for the thickness of the plaster and canvas that is applied before they are positioned.

For treatment of the mould, fixing casting rules, and the particular release agent, etc., see under specific mould.

Casting by the two-gauge method – firstings and seconds

Two containers of plaster are gauged: the firstings, which is applied to the mould first and will form the face of the cast, and the seconds, which is applied to the firstings when it has stiffened sufficiently and in which is incorporated the reinforcement.

Although correctly gauged plaster of Paris has a bowl life of 20 minutes or so, the working of the firstings will cause it to set rapidly. Thus, in all but the smallest of casts, a quantity of SIZE is needed. Enough size is mixed with the seconds to retard the set until the cast is completed.

The required amount of water is placed in the two containers, their size varying with the size of cast. The size water is stirred and a judged amount added to the water in the two containers.

Size is not the only variable controlling the set when casting. Thickly gauged plaster will set more quickly than thinly gauged and, therefore, the plaster should be gauged to the same consistency each time. Plaster splashed hard on to the surface of the mould will also set more quickly than that merely dropped from the brush. This can be made use of when the mould has vertical or near-vertical surfaces. The first splashing application is hard so that it 'picks up' quickly, no set taking place meanwhile in the plaster remaining in the container. A second layer may then be splashed on when the first is stiff enough not to run down and to hold up the second splashing.

Gauging

The plaster should be gauged as described in RUNNING ON A BENCH, and the operation finished by stirring the seconds first and the firstings second. During casting, plaster is carried in a bowl of about 2 litre capacity, held in one hand while application is being carried out with a splash brush held in the other. Before the start of work, a wet splash brush should be shaken and thoroughly worked into the plaster. Surplus water will produce a lighter coloured patch of soft, cheesy plaster on the face. (Fault 1)

Applying firstings

The firstings is first brushed thoroughly all over the face of the mould with the splash brush. Air may be trapped in arrises in the form of bubbles. (Fault 2) To counteract this, the fingers should be run along the arrises, forcing plaster into them and removing the air. Fingermarks should be removed by re-brushing. Excessive brushing should be avoided on shellacked and greased moulds as this will wear off the grease and may cause the cast to stick. (Fault 3) The technique is to spread the plaster efficiently with the brush rather than brush as if one were applying paint.

When the mould is covered with a film of plaster the rest of the firstings is applied by systematically splashing on an even layer which will vary in thickness according to the type of cast. Thickness will normally be between 4 and 7 mm. The hairs of the brush should not be allowed to flick against the mould during splashing as they will wipe off the firstings where they touch. When the mould has been covered the firstings is finished with. The splash brush should be washed out, along with the bowl and gauging container.

The mould is now 'struck off'. This operation entails wiping off all plaster from the strike-offs. On plaster moulds this is best done with a piece of soaked lath. Dry lath will eventually wear off the shellac.

At this point, correctly sized firstings should be picking up and at just the right stage of its set to receive the first layer of canvas. This is now stretched tight over the cast, lowered on to the plaster, and pressed into the top of the firstings with the hands. On large, flat areas it is sometimes pushed in with a gauging trowel as in this way it can be mashed well into the top of the splash texture without being forced through to the face of the cast.

Canvas should be pushed well into all internal angles contained by the mould. If it is allowed to drape across them cobwebs will form. It will be difficult to get the seconds to penetrate these fully and a cavity will exist immediately behind the firstings. These areas will cause the face to cave in when touched. This fault is known as 'shelling'. (Fault 4)

If the canvas is applied while the firstings is too soft it will be pushed through to the face. (Fault 5) The plaster should be allowed to set to the correct consistency or dusted as in casting by the one-gauge method.

The ideal is to have the firstings only stiffened sufficiently to receive the canvas so that the seconds marries completely with it. The two sets should be as close together as possible. If they are too far apart or the firstings is too hard to push the canvas in properly, cockling of the face will result. (Fault 6) If the firstings does happen to be too hard to receive the canvas, seconds may be thoroughly brushed on in an attempt to 'catch' it.

When casting from a pliable mould, care should be taken not to press too hard as the face of the mould will give and the not fully set firstings will crack. The face of the resulting cast will look as though it has been smashed. (Fault 7)

As soon as the canvas is applied, a coat of seconds should be brushed well on to the back to mingle with the canvas and firstings. This may require light or heavy pressure according to the state of the firstings.

Ropes

Any ropes being used should be soaked and placed in position. Ropes are long strips of canvas soaked in plaster and squeezed gently into the form of a rope. They are not used in all casts but may be positioned under timber reinforcements for the following reasons:

1 To prevent the wood cracking the face of the cast (Fault 8), especially with the larger timber sections, such as 25×25 mm, 25×50 mm and upwards.
2 To permit laths to be boxed round the perimeter of thicker casts.
3 To raise a lath to the back of a cast in thick areas where it is to be struck off to a special thickness for fixing purposes, etc.

Laths

The laths are taken from the LATH TANK and pasted. This is done by laying them on the back of the cast and brushing plaster well into all their surfaces. This forces the plaster into the texture and grain, ensuring a good mechanical key. A lath should not rattle about inside a cast as, in this condition, it will not be able to transmit its tensile strength to the structure.

Hollows under flat laths are also troublesome when fixing. On punching home the nail, it will suddenly disappear into the cast, having been offered no resistance. (Fault 9) For this reason, laths that are not positioned on ropes but placed on the back of the cast should be 'puddled' into a line of plaster and not laid on to a nobbly canvas surface. This may be achieved by splashing a general layer of seconds all over the cast before the laths are positioned, also providing a moist back for the second layer of canvas. As the canvas is pressed all over the cast with the hands the plaster will ooze up through the mesh, producing maximum penetration. This is essential if maximum strength is to be gained. A second layer of canvas applied to a dry back cannot be properly impregnated by brushing alone and a cast so produced can be weaker in extreme cases by a factor of up to 5.

The procedure for positioning and covering the laths will vary in different casts according to the scheme of reinforcement. If the laths running both ways are all flat they may be positioned together and covered by turning in and applying the second layer of canvas. If, however, the laths one way are on edge or are on edge both ways (those in one direction being broken between those in the other) they will need to be covered with strips of plaster-soaked canvas wide enough to join them to the cast on both their sides. Therefore, for wooding see under specific type of cast.

In some cases it will be found that more than one layer of canvas is being turned in. One must make sure that all these layers are thoroughly impregnated with plaster and not turned in dry.

On turning in, it is essential that the cast is struck off again. This will remove any firstings that will have oozed on to the strike-offs of the mould as the canvas was pushed into the firstings round the edges, and allow the lath and canvas to be pushed below the level of the strike-offs while the plaster is still soft.

Completing the cast

When all the woods have been positioned and covered with canvas, the cast is brushed in. If the cast has been kept wet, enough plaster will be available on the back to carry out this operation with perhaps just the aid of a bowl of plaster for a cast of some $2.00\,m^2$ in area. All the mesh of the canvas should be filled with plaster, leaving no cobwebs. Once the cast has been brushed in, cobwebs may form round cross-brackets, etc., because of the canvas shrinking on becoming wet. These cobwebs should be re-filled.

At this point, the cast will have a 'brush finish'. Some plasterers prefer a 'splash finish'. This, applied properly, is a light splash providing not much more than a texture on the back of a wet cast. The intention is not to empty on a bucket or so of plaster (as with the firstings) which can increase the weight by up to 50%; this will add practically no extra strength but actually reduce the reinforcement: plaster weight ratio. To thicken the cast deliberately with a splashed finish will both strengthen the panels between the wood lath and, along with a ruled-in 6 mm firstings, help to combat PATTERN STAINING. With a heavier cast of this nature (which may be up to twice the weight of the lightest possible comparable cast) extra reinforcement should be incorporated and may be achieved by placing the wood laths closer together.

To complete the cast its strike-offs are built up. Plaster is first cleaned from the strike-offs of the mould. If the remainder of the seconds in the container is not stiffening to the right consistency, dry plaster may be added to bring it to the correct workability. This plaster is now placed along the laths in the cast to be struck off and the striking off effected by drawing a piece of lath over the strike-offs on the mould, thus accurately transferring the angle and thickness of these strike-offs on to the cast.

As with the firstings, it will be found that correctly sized seconds will set as it is worked so that each stage of the cast will pick up as it is completed, so much so that a fully set cast may be removed from the mould while left-over seconds in the container may still be quite soft.

Some types of cast need extra temporary bracing or lifting-out sticks that are cut off on site immediately before fixing, while others need integral 25×50 mm battens in order that they may be fixed to certain types of fixings. These forms of extra bracing are laid in position across the back of the cast and wadded to the laths in the cast, preferably where two laths cross. If the wad is attached between the laths the weight of the cast on the wad will pull the panel back, leaving a fractured depression on the face and rendering the batten useless as a support as well as damaging the cast.

For the need for lifting-out sticks, integral fixing battens and false braces see under specific type of cast.

Sequence of operations

1 Cut woods and put them to soak in lath tank.
2 Cut canvas.
3 Grease up.
4 Gauge up: measure clean water; add size; add plaster to water; grease hands while plaster soaks; stir plaster with hands to correct consistency.
5 Apply firstings: work brush into plaster; brush plaster into detail; run finger through the arrises; brush out finger marks; splash on an even thickness.
6 Wash out: bowl, brush, main plaster container, hands.
7 Strike off.
8 Apply canvas: press well into firstings.

9 Brush in seconds; splash on a coat so that the laths and canvas may be puddled in.
10 Take laths from tank and paste them.
11 Position laths (with or without ropes).
12 Turn in on perimeter laths and paste if necessary.
13 Strike off.
14 Cover cast with second layer of canvas.
15 Brush in.
16 If appropriate (see text) position laths and cover with canvas until wooding is complete and brush in.
17 Splash back – optional.
18 Wash out splash brush.
19 Strike off; build up strike-offs.
20 Position any lifting-out sticks or false brackets and wad up.

N.B. On a large cast, which has been properly sized, seconds agitated in the hairs of the splash brush during working will cause it to pick up and set. As the brush stiffens it should, of course, be washed out but is normally washed out as a matter of routine at points 10 and 15, otherwise it may set solid during the periods when it is not in use.

Faults in casting

1 Water on the face.
2 Air holes.
3 Sticking in the mould (usually only with plaster moulds) and causing arrises or patches of firstings to be left behind.
4 Shelling.
5 Canvas on the face.
6 Cockling.
7 Crazed firstings.
8 Laths cracking the face.
9 Loose laths caving in during fixing.
10 Thick grease on the face.
11 Strike-offs not made up properly – too high, lumpy, etc.

Casting by the one-gauge method

The method is the same as that for two-gauge except for amendments to the first coat of plaster. One of two procedures may be adopted: (a) the plaster is a true one-gauge – all mixed together in one container to a thicker consistency; (b) the plaster is gauged to a normal consistency in one container. Sufficient plaster to cover the mould with the first coat is taken out and stiffened with dry plaster. This, of course, has the effect of reducing the plaster : size ratio.

In either case the plaster is applied as firstings and prepared to receive the canvas by 'dusting'. This process is intended to prevent the canvas being worked through to the face of the cast. The 'dust' is dry plaster sprinkled evenly over the cast by shaking it from a dust bag which may be made by gathering the corners of a square of one or two thicknesses of canvas. Care should be taken to avoid excessive dusting. A layer of uncombined dust will act as a separating layer. During handling and nailing this can result in chunks of firstings splitting away.

Faults

With one-gauge work, shelling and cockling should not occur. The other listed faults may occur and splitting away of the face due to excessive dusting is an additional hazard.

CASTING IN KEENE'S CEMENT

Keene's cement is not currently commercially available but this entry may be of use in restoration work. Fibrous plaster casts are made in Keene's cement when strength and resistance to knocks is required, e.g. casts in vulnerable positions, stock moulds, and moulds for casting in sand and cement. The technique is the same as for CASTING IN FIBROUS PLASTER, one-gauge, but provision must be made for dealing with the situation where no set is taking place in the fine, silky plaster from which the gauging water will stream constantly.

The first coat is applied as a paste as thick as can be brushed into the detail and flopped on with a splash brush. This coat is dusted and the canvas applied. The cast is completed with plaster as thick as will still properly impregnate the canvas. DUSTING is advisable to soak up moisture after every brushing. A cast kept as dry as possible in this way can have its strike-offs built up and ruled off immediately before the plaster picks up. Normally, due to the slow set, only one cast can be taken from a mould in a day.

CAVETTO

See MOULDING, SECTIONS.

CEILING, BARREL

See VAULTING, barrel.

CEILING, RIBBED

A vaulted ceiling containing mouldings taking the form of ribs.

CEILING, SUSPENDED

A suspended ceiling is one that is suspended from the structure of a building by means of a grid. It can be made up of fibrous plasterwork or solid plastering on metal lath. See GRID, FOR FIBROUS AND SOLID PLASTERWORK.

CEMENT CAST See CASTING IN CEMENT

CEMENT, HAZARDS See APPENDIX 1, P.279.

CEMENT, HIGH-ALUMINA

High-alumina cement is manufactured by melting and fusing a mixture of limestone and bauxite (aluminium ore) in a furnace. The molten mass is allowed to cool and then ground to a fine powder. It is very different in composition and properties from Portland cement and is comparatively slow-setting but rapid-hardening. Where possible, it should be kept moist for at least 24 hours from the time it begins to harden. Hydration is generally complete after a few days and there is no progressive gain in strength.

CEMENT, PORTLAND

The raw materials of Portland cement are clay plus chalk or limestone. There are two basic methods of production: the wet process, in which the materials are combined in a slurry, and the dry process which combines them in powder form. In both processes the materials are then fed to a rotary kiln and eventually reach a temperature of 1400°C. Various chemical changes take place in the kiln and the resulting clinker is cooled and conveyed to ball mills. Before crushing, raw gypsum is added in a proportion of 4 to 7%; this is to retard the set. The powder thus produced is Ordinary Portland Cement (OPC).

CEMENT, PORTLAND, RAPID HARDENING

As for CEMENT, PORTLAND, but more finely ground than OPC. It should not be regarded as a quick-setting cement as the term 'rapid hardening' refers to the rate of increase in strength and not the setting time.

CEMENT, ROMAN

A hydraulic cement made from a type of clay found in some rivers. It was usually brown in colour and resembled the mortar used by the Romans. It was made by burning and crushing the clay. At the time of the Great Exhibition of 1851 its strength was deemed to be about one-third that of Portland cement. Mixed with sand, it was used for some external finishes. For casting it was general practice to use it either neat or with just one handful of sand per bucket. It set fairly quickly and was often used in a way similar to CASTING IN FIBROUS PLASTER, one-gauge work.

CHALK LINE

See LINE, STRUCK OR SNAPPED.

CHAMFER

A splayed external angle that is formed, if plain in section, or run, if moulded.

CHANNEL

The shape of the section used to describe a metal member. See CEILING, SUSPENDED.

CHATTERING

The juddering of the stock of a running mould. This causes the formation of a continuous series of flutes or grooves some 1 to 2 mm wide across the section being run. Accepting that the running mould is horsed properly, it is caused by the profile being required to cut off too much expanding plaster at one time. This may be brought about by (a) excessive expansion caused by too great a volume of plaster being used to bring out the section; (b) inexpert running on large sections, even when a muffle is used, e.g. too much time elapsing between runs or a change of pressure on the running mould from one run to the next – this will require the profile to cut off differing amounts of plaster from the run and, to ensure against this, each person, no matter how many are working the mould, should take hold of the running mould in exactly the same place and push or pull in exactly the same way, exerting as far as possible the same force each time the mould is run. See also CORE; MUFFLE; RUNNING ON A BENCH; MOULD, RUNNING/HORSE.

Course of action

Allow the running mould to ride by applying less downward pressure: the feel of the running mould will indicate the amount of pressure. The profile should still scrape the surface of the section – if it is allowed to lift, gathering on will result. Run the mould over only once or twice more, feeding it with soft plaster. Alternatively try running the mould over backwards, cutting off the higher parts of the chatters. Chatters on a finished moulding may be cleaned off with a joint rule.

CHEEK

The vertical side of a BEAM CASE, etc.

CHEESY

A consistency of plaster at one stage of its set from a creamy liquid. The texture is that of a block of butter.

CLAY

Modelling clay has many uses in fibrous plastering, e.g. taking SQUEEZES, forming CORES for clay-case moulds, for walls and fences in flood and skin moulding, sealing pour funnels to cases, and plugging air holes in case moulds. Clay in use will dry, shrink and crack; therefore any clay surface that is not to be worked on immediately must be prevented from losing its water content. This can be done by a covering of wet paper, canvas or polythene, either singly or in combination.

After use the clay will be too hard for further use and will contain foreign matter. To make it suitable for re-use it is made into a slurry by mixing with clean water and then passed through a sieve. It is then poured into shallow trays and allowed to lose moisture until it has body enough to be made into a mound. Pouring the slurry on to a thick, dry slab of plaster will aid in the initial extraction of water.

Further drying is aided by grooving the surface (and thus increasing its area) by hitting it with a timber batten held on edge. When it has reached the correct consistency it is stored in adequately airtight galvanised bins until needed again. If drying out occurs, the clay may be kept damp by covering with wet canvas and/or polythene, within the bin.

Beating

When an even thickness of clay is required for use as walls and fences or cores for case moulds, a CLAY BOARD is used to permit the clay to be beaten into even sheets of a controlled thickness. The board is first dusted with french chalk in order to prevent the clay from sticking to it. The correct amount of clay is placed between the rules and kneaded together by beating it with a wooden rule (used on edge to obtain a better point load). The clay is beaten until a small uniform surplus exists just above the thickness of the rules. It is then cut level by traversing off the side rules. The sheet of clay may then be rolled back in the form of a swiss roll, the french chalk picked up from the board on its underside preventing it from sticking to itself.

For clay as an impurity see SAND.

CLAY BIN

A galvanised bin with a close-fitting lid in which clay of the correct consistency for use is stored.

CLAY BOARD

A clay board is used in the beating of clay as described under CLAY, beating. It can vary from about 400 to 600 mm long but the width is usually about 300 mm. It may be made from plywood, blockboard or planks with thickness rules down the long sides.

CLAY SQUEEZE

See under REPAIR AND RESTORATION.

CLAY TILES (POTS)

See BACKGROUND.

CLAY WASH

A release agent used for plaster in waste moulding.

CLINKER BLOCK

See BACKGROUND.

COARSE STUFF

See LIME.

COBWEB

See CASTING IN FIBROUS PLASTER, general method.

COCKLING

See CASTING IN FIBROUS PLASTER, general method.

COFFER

A deep, sunken panel in a soffit, dome, ceiling, etc., formed by mouldings or beams.

COLD FLOW

See COLD-POUR COMPOUND

COLD-POUR COMPOUND

The material is a liquid polysulphide compound which sets to a flexible rubber when mixed with a curing agent. It is available in varying degrees of flexibility and the curing time can be controlled by the choice of curative. A great advantage over other flexible moulding materials is that, because no heat is present during the cure, none of the imperfections caused by the sudden heating of the face of a model occur. It is extremely penetrative and reproduces such fine detail as the very texture of carved coarse plaster. This property calls for sealing butting joins between cases and the join between case and moulding grounds in case moulding – an unnecessary measure with POLYVINYL CHLORIDE and GELATINE, provided the joint is close-fitting.

Seasoning

In general, damp surfaces are not detrimental to curing and, therefore, models of clay and damp plaster may be used as well as plasticine, metal, wood, glass fibre and most plastics. Although seasoning is not required by the compound, a thin coat of the manufacturer's release agent or petroleum jelly thinned with an equal quantity of paraffin (applied to all but clay and plasticine) will aid in parting. The compound will stick to itself, however, and if it will come into contact with already cured parts of a mould – as in back-and-front and multi-piece clay-case moulding – the cured surface should be treated with a silicon, polyvinyl alcohol, or the manufacturer's release agent.

All forms of plaster case containing cold-pour moulds should be sealed with shellac to prevent the porous plaster drawing the plasticiser from the compound. Manufacturers' instructions should be consulted regarding the suitability of their various products for particular models, casting materials and for mixing and pouring procedures.

As with all catalysed materials, both parts should be thoroughly mixed together to obtain a uniform, complete cure. Pouring the mix slowly will cause the air bubbles introduced during mixing to burst as the thin film flows over the lip of the mixing container. Although no heat is given off by the compound, the ambient temperature will affect the speed of curing.

To cast from (plaster and cement mixes)

The mould is given a thin coat of petroleum jelly or the manufacturer's release agent.

To cast from (glass fibre : epoxy or polyester resins)

The mould is protected from resin attack by coating it with a polyvinyl alcohol release agent before each cast. This is applied by brush or spray after first washing the mould with a warm 1 to 2% solution of detergent and allowing it to dry. The release agent should be allowed to dry and care should be taken to avoid cracking it. This will happen if the mould is depressed or bent. Casts producing a low EXOTHERM may be taken from moulds of the standard compound and special compounds are available for use with reinforced plastics and urethane foam.

Stock moulds

The material is subject to 'cold flow'. Therefore, moulds being stored are best kept to shape by the model or a cast, suitably sealed with shellac, being left in them.

Types of mould possible

Flood; skin; clay-case and back-and-front; run-case; insertion. See MOULD, REVERSE.

COLD-POUR METAL

See GLASS FIBRE.

COLLAR

See COLUMN, solid.

COLONNADE

A range or series of columns.

COLUMN

Fibrous

The model for small plain columns may be turned up in a TURNING box and moulded as for BALUSTERS.

Reverse moulds – caps

Enriched caps, along with abacuses, are clay-case moulded in two halves and cast separately from the column. The enrichment is usually modelled in clay on a bell spun in plaster. Although the height of the shaft mould can be arranged so that the case mould can be placed against the end and the cap cast on the shaft, a separate cap can be useful in adjusting the height of the unit slightly to suit fixings. Deep undercut will need special provisions, e.g. loose-piece cheeks in the case round volutes; the tips of acanthus leaves omitted from the model to be moulded and cast separately – dowels are left projecting from the tips to assist in fixing to the cap. Plain-moulded caps, for such orders as Doric and Tuscan, can be turned up with the reverse mould for the shaft.

Reverse moulds – bases

These are usually plain moulded and turned up with the shaft in the column box. Small pieces of enrichment may be planted on afterwards, otherwise it will have to be clay-case moulded.

Reverse moulds – shafts

Reverse moulds for plain parallel shafts may be run off in half with a running mould, the spun reverse moulds for a plain moulded base and cap being joined to it so that the column is cast in one. As parallel fluted shafts are undercut, they are best produced as entasised shafts.

Reverse moulds – entasised shafts

Reverse moulds for these shafts are turned up in half in a column box along with plain-moulded caps and bases. Flutes are produced by reverse moulds of the flute lying in a rebated bed in the shaft. Suitable rebates in the bed should be provided to take straps to tie the flute to the cast so that they are removed with it and not left loose to fall back (Fig. 22).

Although traditionally a column box is made up with two long sides and two ends, on small columns, when a core is being used, only the ends to carry the pivots are required. In either case the ends are best removed after turning to make the striking off of the base and cap easier. Special column irons are not necessary: pivots can be made by drilling an iron bar to take a rod or by using wooden blocks to hold rods in grooves in the ends of the box. Metal stops are needed, however, to prevent lateral movement.

The shaft is turned up with an entasised feather-edged rule, to the ends of which are attached the metal profiles for the plain-moulded cap and base. Handles are fixed to the rule, as shown. The plaster is put into the box and the section built up by revolving the profile (forcing it down the far side and, taking hold of the handles, pulling it up at the near side). It is propped clear with a piece of lath for cleaning.

The first run is muffled with a 6 mm lath on the

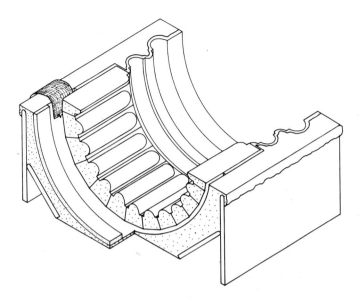

Fig. 22. Column (fibrous): the base of a reverse mould containing flutes

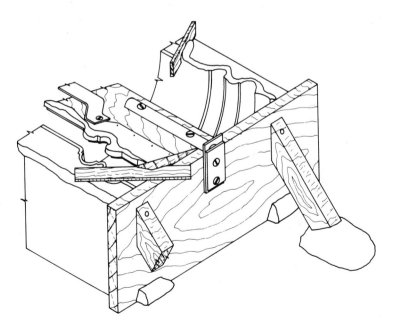

Fig. 23. Column (fibrous): turning up the finish to a reverse mould

shaft and a conventional muffle on the two metal profiles. During the coring out the two longitudinal strike-offs forming half the column are accurately ruled in off the two ends with a straight edge. The strike-offs of columns longer than 3 m may best be ruled in from the side rules of a column box. These strike-offs are shellacked and greased, and used to strike off the finished run. The muffles are removed, the core keyed and running begun. It is expedient to confine the plaster to the metal profiles and run up these sections until only two or three runs are needed to produce the finish. At this stage the plaster is applied to the shaft as well as the ends. This will provide a fully built up shaft and avoid the excessive chattering and gathering on due to the plaster expanding in the shaft. The sides are struck off from the longitudinal strike-offs (Fig. 23).

As with all surfaces formed with a wooden profile, the texture of the shaft will be woolly and is brought to a smooth finish by a creamy gauge of plaster worked in with a joint rule or piece of busk as in topping plaster BENCHES.

Flutes

A profile to run the mould of a flute is cut. In the case of a diminished column, it is cut to the bottom section and the running rules are fixed to the running mould as it follows the setting out of the flute (Fig. 22 (a)).

The flat back obtained by running on the bench is quite adequate and there is little point in running in the column bed or on a similar bed formed by the entasis rule. To obtain the diminish, the width of the flute is taken from the plan and put into the elevation. The flute mould is run off and its ends returned with a gouge or gauge. A mould is taken of the flute mould and the required number of flute moulds cast from it. Where there are fillets the joint is made in the middle of each fillet but in the absence of fillets the joint is made in the middle of a flute. However, the two halves of the column are joined down the middle of a flute and the series of flutes should therefore finish with half a flute on each side. When a reed is present in the bottom third of the

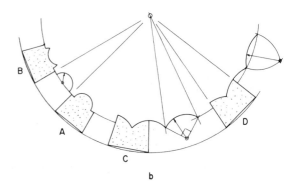

Fig. 24. Column (fibrous):
(a) setting out the diminish
of a flute
(b) sections of the reverse
moulds for flutes

a

b

flute this portion may be run by superimposing the convex plate on the profile of the flute. This is then joined to the top portion and the mould taken. Figure 24 (b) shows the moulds in position against flutes: *A* semi-circular with fillets; *B* semi-circular with a reed; *C* and *D* segmental with arrises – *C* quarter-circle and *D* equilateral.

Casting

Laths round small columns are usually bruised and flat while those in large columns are placed on edge, broken to fit between the longitudinal ones. The latter are on edge, the number depending on the size of the column. False columns that are joined together and made good before leaving the shop have the strike-off laths flat, thus providing a key when wadding up the joint (Fig. 25). Lifting-out sticks are needed so that the casts may be removed from the mould by a cramp (Fig. 26). Occasionally, columns are used as permanent shuttering – fixed and filled with concrete. These will need extra re-inforcement.

Fixing

The halves of a false column are held in two padded cradles and tourniqued together. (Fig. 25). The joint is pasted and a continuous wad poked along its entire length with a batten and brushed well in. When both joints have been wadded they are made good with a flexible piece of busk. The capital can also be wadded in position and made good. The plinth is usually in wood and is united with the column on site.

Fixing on-site

Where timber fixings are provided they are utilised in the normal way. Where the halves are fixed to

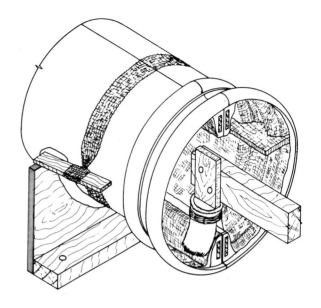

Fig. 25. Column (fibrous): joining the halves of a false column

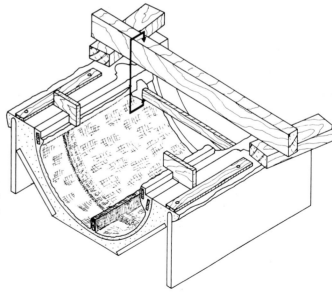

Fig. 26. Column (fibrous): the cast and its removal from the mould

steel or concrete stanchions wads may be wedged between structure and cast. Blocks and wedges are used to hold the first half in position while it is plumbed down with offset gauges cut to suit the entasis (Fig. 27). The cast can then be securely wadded at intervals down both sides. When the wads have set, the second half can be offered on and fixed to the first by nailing or screwing through the laths in the lapped joints. In the absence of lapped joints, blocks of wood can be positioned in the fixing wads to which the casts may be fixed. On plain columns extra security is obtained by tourniqueing round the column with galvanised tie wire countersunk in a saw-cut groove. Separate caps cast in two halves are now wadded in position on the column.

Stopping

On plain columns flexible busk is used to make good joints, nail and screw holes and countersunk tourniques. A metal disc to fit the flutes is used in conjunction with a short joint rule on fluted columns.

Spiral columns

These are best cast from a cast two-piece mould, incorporating base and cap where possible, as above. The model of the shaft is turned up in a TURNING box. Usually only one repeat of the spiral is turned. This is piece moulded and enough repeats cast to make up the shaft.

Solid

Backgrounds for plastered columns may vary from rough-cut brickwork or circular cast concrete to steel joists, cradled and metal lathed. When preparing to form a circular column in solid, an accurate measurement of this background should be taken and the minimum thickness of plaster to be applied, both at the top and at the bottom, noted. Taking the minimum thickness as a guide, a metal profile should be cut to spin a plain plaster band, the

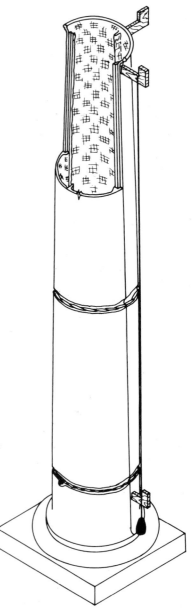

Fig. 27. Column (fibrous): plumbing a diminished column

width of which will be the minimum thickness. The depth can vary between 50 and 60 mm and the mould horsed up for spinning on a bench or spot board. A gigstick is then fixed to the running mould and the measurement from the outside member to the centre point or pivot will be the correct radius for the column size.

In the case of a diminishing column, two radii will be used, one for the bottom of the column and one for the top. A complete circle should be run, but the completed collars will have to be cut into two halves for fixing purposes.

Plain straight column in situ

Having spun the two plaster collars, these may now be fixed in position (Fig. 28(a)). If capitals are to be fixed when the columns are completed, the top collar should be fixed so that the bottom of the collar will coincide with the bottom of the capital cast. Once this has been fixed, usually by bedding in neat Class A plaster, the base collars can be held in position whilst a plumb bob is dropped down and the bed prepared to receive them. Should cast bases be required here, the top member of the collar would be at the same height as the top member of the base. For the best results the top collars should be marked at four points, all at right angles to each other, then the lower collar can be plumbed down from these four points. For large columns use a plumb bob and gauges; for smaller columns a parallel rule and spirit level will suffice.

A straight floating rule is then prepared by cutting to the size of the column and fixing two slippers that have been cut on one face to the column curve, at either end of the rule. These will bear on the two plaster collars and, if the cutting is precise, will run round the column easily and keep the rule in an upright position. A muffle may be fixed to the rule so that the backing will be cut back to the required depth for the finishing coat.

Apply the backing coat evenly, and gradually

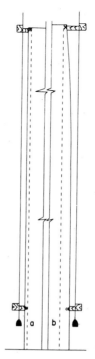

Fig. 28. Column (solid): plumbing collars to parallel and diminished columns

bring it out to the required thickness. The rule is passed round the column and, when the floating is in contact with the rule at every point, it can be keyed to receive the finish. This will mean devil floating for gypsum plaster internal columns and comb scratching for cement-and-sand plastered external columns.

Clean the collars, remove the lath or muffle from the rule and, when the backing coat has set, the finish can be applied. For a gypsum plaster finish, lay on sufficient plaster horizontally with the trowel and pass the rule round the column. Repeat until a perfect shape is obtained, then lay down with a skimming float and finally a steel trowel. Constantly check the shape of the column with the rule during these operations.

Depending upon the type of gypsum plaster used

the actual finish may be obtained by trowelling twice with a laying trowel and a little water or, if Keene's cement is used, by scouring and polishing.

Should the column be in cement and sand, build out again till a perfect shape has been obtained by the rule and rub up gently with a wood float as in external cement plainface. (See FINISH/RENDERING, EXTERNAL.)

The importance of the collar positions can now be seen. When removed, a rim will be left, on which the cap may be rested whilst being fixed. The same applies to the base. For external columns, the plaster collars must be removed, the cement-cast cap will then sit perfectly on the lip. Care should be taken in the removal operation so that the column is not damaged in any way. Only the shaft is formed in solid by application of plaster to the background. The caps and bases, whether in plaster or cement, are cast from reverse moulds that have been previously prepared.

In the case of external columns the casting is often carried out on site. The same mix that is being used for the shaft can then be applied. This should result in the caps and bases being the same colour as the shafts.

Plain diminished and entasised

The procedure is the same as for straight columns, with one exception: the floating rule must be cut to the entasis of the column. (See Fig. 28(b).)

As mentioned previously, the two plaster collars will have been run to two different centres. The plumbing down should be done in the same way as for straight columns, but the top collar will be back behind the bottom collar. Therefore a pair of purpose-made gauges will be necessary, the top gauge taking into account the degree of diminish.

Great care must be taken to ensure that the entasis rule is kept perfectly upright when traversing round the column. Any deviation from the vertical will mean a change in the shape of the entasis.

This should make obvious the necessity for the two curved slippers, one at either end of the rule.

The finishing once again should be applied horizontally and a good flexible laying trowel will be necessary for the trowelling up. For a cement and sand finish, the final coat will be finished by a wooden float.

Fluted diminished column, internal

Fluted columns in solid have for many years been considered one of the most difficult operations in solid plastering. In the past, many writers have given as the best method for forming these – either in gypsum plaster or cement plaster – the plaster-collar method. This entails considerable additional skill in obtaining a perfectly smooth finish; therefore, wherever possible, we feel that the following method is more suitable for gypsum plaster fluted columns.

Measure the overall dimension of the column, top and bottom, and run two plain plaster collars as before, keeping the floating line at least 12 mm back behind the flute depth line. Plumb these in position and prepare an entasis rule complete with the two slippers. Finally float the columns to perfection and key. The next step is to set out another bench-type metal profile, consisting of one half-flute, one full fillet and one half-flute, and horse this as for hinged moulds.

This mould should, at the base of each half-flute, have a 12 mm deep member, as allowed for in the floating; also these members should be filed to an undercut. On a bench or long, supported scaffold board, prepare a running base, well over the width of the prepared hinged mould, to the same shape as the column curve, and to the full length of the column. This can be done by using a section of the two collars used when floating, plus the entasis rule.

When this is complete, shellac and snap a centre line throughout its length. At one end, set out the full width of the running mould, and at the other end, the diminished width, equidistant to the centre line. All the intermediate setting out should be checked by the entasis rule. When satisfied that all is well, fix two good running rules, one either side of the moulding for the outsides of the slippers to bear against. Then fix two smaller rules for the inside edges of the slippers to run against, small parts of the wood stock will have to be cut away to allow for the inner rules. They are necessary to keep the running mould in position as the two outer rules close up, forcing the hinges to close and the mould to get smaller. Grease well and run a full length off, reinforced with canvas.

Meanwhile it may have been possible for another craftsman to have set out accurately on the column backing coat the position for bedding up when the moulding has been run. This must be accurately done all round the column, so that each run can be bedded up immediately it has set. To move it, the best way is to have a firm straight board the same length as the moulding. Lay this on the bench beside it. Remove the moulding from the bench, key back and turn on edge as one lifts it and put it on the board.

It may be necessary to tie it to the board, then lift them both till they are upright on the column. The area where it is to be bedded should be treated with both water and whatever adhesive is available, from plaster to PVA. Then untie the cord and slide the fluting off the board into its correct position, pushing it well back at the same time. Trim off all surplus material from the edges, especially at the joints. Repeat the entire operation and it will be easy to see just why the 12 mm thickness members were made undercut. For this means that the edges at the flute centres can be brought together precisely. Normal members would have produced a gap through meeting at the bottom edge.

Having run one length, some plasterers may prefer to cast a reverse mould from this run, in which case the flutes would then be cast, instead of being run. Advantages are that each cast would probably be better reinforced, but care would have to be taken to see that the strike-offs were perfect. A disadvantage is that one could not cast the undercut members, so in this case they would have to be made straight and chewed off by surform or block plane when set.

Fluted column, diminished, external, in cement and sand

Once again, personal preference is for the plaster-collar method, though, in the past, at least one other method has been proved fairly effective. To produce the collars, it will be necessary to prepare two semi-circular half-column running moulds. The fluting must be included and the setting out and filing must be done accurately. The diameter of one will be to the column size just above the base, the other, the diameter just below the capital. Muffle the two running moulds to a thickness of 10 to 12 mm below the flute depth line and run off two cores to these sizes. Remove the muffles and, with the smaller mould, run over the shellacked and greased core enough plaster moulding to produce two 75 mm collars. With the larger running mould, run sufficient for four 75 mm collars.

The reason for this is that most fluted classical columns are only entasised for two-thirds of their height, the lower third being perfectly straight. Fix the top two half-collars in position all round the column and then, using gauges specially prepared to allow for the diminishing, fix two of the larger collars a third of the way up. When this has been completed, plumb the two bottom collars down from the central pair.

This whole operation should be taken carefully, as not only must the collars be planted correctly to the column shape, but the flutes must also be perfectly placed, one above the other. Some people will prefer to fix top and bottom collars, then line in the central ones. This is perfectly all right. In fact when working on a large column that is diminished and entasised from top to bottom, a third set of collars, run off to the column size centrally is always fixed last.

Three rules will be required for the floating and finishing: the normal, large outside entasis rule, complete with a slipper at both ends that will run smoothly across the fillet faces; a feather-edge rule cut to the flute shape and size; and a straight feather-edge rule (Fig. 29). A similar set will be required for the bottom third of the column – this time, all three will be straight.

The floating may be brought out to a muffle attached to the two shaped rules or it may be roughed out and gauged back. Obviously the better way is to use the correct rules so that the thickness of the finishing coat will be constant over the entire area. There are varying opinions on the practical procedure, and these will again vary according to column size. However, in general, the gradual building out in thin coats to the whole column makes for a more stable job. Pass the entasis floating rule completely round the column until the fillets are more or less complete. Then using the fluting feather-edge rule, cut out the flutes one by one. The straight rule is for holding tight against the adjacent fillet while the fluting rule is being slowly pivoted. This should prevent it from being pushed off. The fluting rule must be kept sharp and clean so as to cut into the material.

When the backing has set, remove all muffles and finish in more or less the same way. Great care must be taken to see that the shape of both the fillets and flutes are built out perfectly – bring the work along so that everywhere is approximately at the same stage. This means of course the bottom third of the column as well.

The final texture will be obtained by using small floats of all shapes and sizes to the whole column. Returns at the flute ends are worked in by hand. The best way to do this is to prepare a reverse mould of a perfect flute end in either wood or plaster. Then having marked all round the column the position

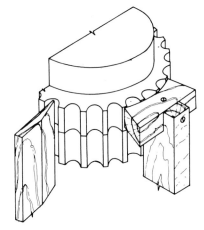

Fig. 29. Column (solid): ruling in from a fluted collar

that these will be, one should hold the mould in the flute, slide it up till it is in its correct position and fill in the excess fluting with cement and sand, and finish with a wood float.

For columns that have no fillets, the procedure is the same. However, it may help if the collars are in hard plaster so that when the horsed entasis rule passes round it will run across the sharp arrises without breaking them. It is also a much more difficult job to work up to the sharp arrises in cement and sand than to good wide fillets.

Provided note was taken of the positioning of the collars as recommended earlier, on their removal they will supply a perfect position for both cap and base. The central ones should be carefully removed and the gap made good while the column is still green.

A refinement of the collar method was tried a few years ago and proved reasonably successful on both small columns and diminished pilasters. A full-length reverse mould was cast of a flute, without the return ends. This was reinforced with angle iron and had several twisted wire handles projecting from the back of the cast. The fillets of the column were formed fully by the horsed entasis rule, and the back of each flute was floated out just full of its requirements.

At this stage the plaster reverse mould was forced into the flute and moved very slightly in an up-and-down manner. This action was continued until the plaster mould nestled snugly into the top and bottom fluted collars. The action taken for its removal was to slide it gently downwards. Room had been left for this. As the flutes diminish at the top, the mould came away quite easily, leaving a near perfectly formed flute and good fillet arrises. These had been worked in accurately to the edge of the plaster mould. Time has not permitted further test on very large columns, but it is worth a try.

COLUMN BENCH

See BENCH, plaster; COLUMN.

COMPOSITE BACKGROUND

See BACKGROUND.

COMPOSITION (COMPO)

A casting material and technique first introduced into Britain in the late 1700s. It is still used today, mainly for making ornamental mirror and picture frames and low-relief ornament. To produce the enrichment to dress run plaster models.

To prepare composition, two pans are required. Into the first, 3 litres of water are poured and to this is added 8 kg of best quality Scotch glue. This is dissolved thoroughly and should be continually stirred. Into the second pan, 2 litres of linseed oil plus 4 kg of ground resin are placed, heated and stirred. When both are dissolved, the contents of the glue pot are added to the resin pot and the two thoroughly mixed. To this, sifted whiting is added until the compo acquires the consistency of a thin dough. It is then turned out on to a clean whiting-dusted board and the kneading and the addition of whiting should continue till the desired consistency has been obtained. This will vary with the shape, size and depth of the mould.

The reverse mould from which the compo will be cast will be one of four types – a reverse carved box-wood mould, a sulphur mould, a hard plaster mould or a reinforced plastic mould. All of these will require a light application of linseed oil, and the warm compo is rolled smooth and cut into strips then hard pressed into the mould. A wet board, slightly larger in surface area to the reverse mould is then laid over the cast and the whole placed into a press.

Pressure is then gently applied until the compo has been pressed into every corner of the mould. The pressure will cause the wet board and the compo to unite, so that when removed from the press the board can be lifted and the cast will be stuck to it, usually on a flat ground of surplus compo. The cast is then trimmed to its correct depth with a wide knife, and the surplus composition returned to the pot for re-heating.

When fixing, the back of the cast may be lightly softened by steam. One method for this is to place a large joint rule flat over the rims of a bowl filled with hot water. When 'tacky', place the cast gently in position and pin with fine panel pins.

The ends of the casts must be accurately trimmed when soft so that when planting or bedding a continual line of enrichment no stopping will be necessary.

It is now possible to buy casts of this type of ornament from shops. The casting material is plastic and these may be fixed with an adhesive made from a similar material.

CONCRETE

See BACKGROUND.

CONSOLE

A rectangular block, enriched or plain moulded, acting as a support to a feature. It is fixed with its projection exceeding its depth. For model, see BRACKETS.

CORBEL

A block on a wall made up from a moulding to act as a support for a feature such as an ARCH.

CORE

Fibrous

Type 1, where the core becomes an integral part of the run

An infill of coarse material used to fill out the bulk of a section when running. The core for bench running may be formed in the following ways:

1 Built up with pieces of set plaster (old casts, mouldings, rubble, plasterboard and so forth.
2 In the form of a DRUM with RIBS or templates, lathed over and covered with plaster-soaked canvas.
3 Built of bricks, blocks or similar materials, covered and protected with paper so that they are neither stuck together nor dirtied by the plaster and may, therefore, be used an indefinite number of times.

4 Run-in plaster of Paris with a MUFFLE and well keyed. It may be necessary to core out the muffled run itself, using one of the three methods described above.

In the following instances, it may be necessary to use a core to aid running:

1 Where the volume of the section to be run is large enough for the expansion of the setting plaster to cause CHATTERING or even prevent the running mould from running altogether.
2 Where the moulding contains a section that will prove difficult to build up during the time the plaster takes to set, e.g. thin upstands, overhanging and undercut sections.
3 In conjunction with a metal plate to form a bed for a loose-piece.

A core is additionally useful in that it provides suction, either green suction from a freshly run core or complete suction from dry casts or rubble. This will allow the section to be built up before the plaster stiffens due to its set. Although it is possible to run large cores with neat plaster, a quantity of SIZE may be added – a good finish on the run is not required. As the core hardens it should be well keyed every 30 mm by diagonal undercut grooves cut with a chisel. The plaster of the finishing run should be well rubbed in, for after a matter of seconds the undercoat will have sucked the water from the plaster, making it too stiff to be forced into the key.

Type 2

Here, the core itself has been run with a template and shellacked and greased in order that the moulding run over it may be removed, giving it a hollow or shaped back. See RUN CAST; MOULD, REVERSE, run case.

Type 3

A core off which the case is cast in clay-case or run-case moulding. See MOULD, REVERSE.

Solid running

Cores or coring out for a solid-run cornice will usually take one of the following forms:

1 A 50 : 50 mix of plaster and sand or plaster and coarse stuff laid into the angle of the wall and ceiling well behind the finished run. This method is used on medium-sized cornices.
2 Successive coats of a similar material applied to brackets or bracketing and eventually run out to a muffle on larger cornices.
3 Narrow strips of plasterboard fixed across the wall ceiling angle and covered with Class B board finish plaster for small and medium-sized cornices.
4 In the case of external moulded cornices in Portland cement and sand a core must be brick corbelling, concrete or stonework built out behind the place where the cornice is going. This would be spatterdashed and rendered with successive coats of a mix similar to that of the finished cornice. Alternatively, metal brackets could be used, wired with EML and rendered. In both cases the external cornice would be built out to a muffle.

CORK

See BACKGROUND.

CORNER MESH

See BEAD.

CORNICE

A cornice is the uppermost moulding of the three that form the classical entablature and allied features, such as PEDIMENTS, OVERDOORS, mantelpieces, etc. Also used internally as the moulding at the junction between wall and ceiling and externally at the top of a building or just below the parapet.

Fibrous

Straight – plain moulded

The reverse mould, the profile for running the reverse mould (a), the model, and the profile for running the model (b) can all be drawn in round the section of the moulding as shown in Fig. 30. See also MOULD, REVERSE, run loose-piece, run slip-piece.

Curved – plain moulded

The reverse mould from which a curved cornice is to be cast must be run in projection with its section in position relative to the curve. The nib of such a running mould is apt to cause CHATTERING. The unwanted back of the reverse mould is, therefore, not run up and the section is finished at the top strike-off. Figure 31 shows the set-up for a cornice to a curved ceiling. Such a section being run other than with a gigstick (e.g. with a peg mould, eccentric rule, etc.) will need to have its nib supported by a rail. The set-up for a cornice to a curved wall is shown in Fig. 32.

Enriched

See MOULD, REVERSE, clay case, run case and insertion.

Casting

As for CASTING IN FIBROUS PLASTER with the following variations: The cornice is reinforced as for a long, narrow cast. Each strike-off will contain a wood lath, 3 or 6 mm, and a rope if there is sufficient

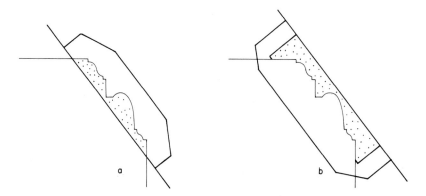

Fig. 30. Cornice (fibrous): (a) profile to run a reverse mould, (b) profile to run a model

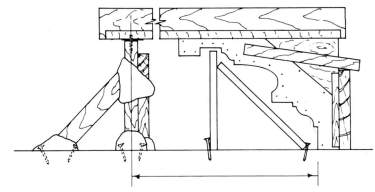

Fig. 31. Cornice, curved (fibrous): spinning the reverse mould for a vaulted ceiling

room. Long laths that need joining have their ends butted with a short piece of lath placed on top or alongside the join to reinforce it. Where ropes are not used, a method of getting three layers of canvas under the lath in the strike-off is shown in Fig. 33. Extra longitudinal laths on edge 150 mm apart are used in casts that are 300 mm wide or over. Cross-brackets are placed every 300 mm or so along the length and may be covered by the turning-in or by separate soaked laps.

Although it is practice to use soaked laps as brackets across the backs of small sections, it is better to include wood lath. Where room behind the cornice allows, the brackets are best placed on edge, either as one piece or with knuckled joints, accord-ing to the section. Laths placed flat are either straight or bruised to follow the section. Large sections consisting mainly of a single curve or straight area are provided with rebates at the ends to allow the joints between the casts to be reinforced with scrim.

Fixing

The fixings illustrated in Fig. 34 are not confined to the ceiling or wall but each may be used in either position. (a) fixing by plugging through finished plasterwork into the background and nailing or screwing through finished plasterwork to timber;

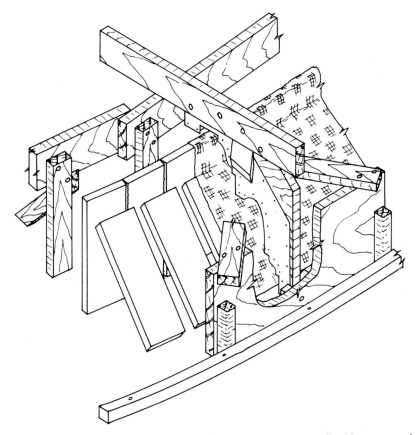

Fig. 32. Cornice, curved (fibrous): running the reverse mould for a curved wall with a peg mould and a rail

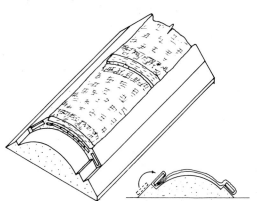

Fig. 33. Cornice (fibrous): run reverse mould and
cast of a cove

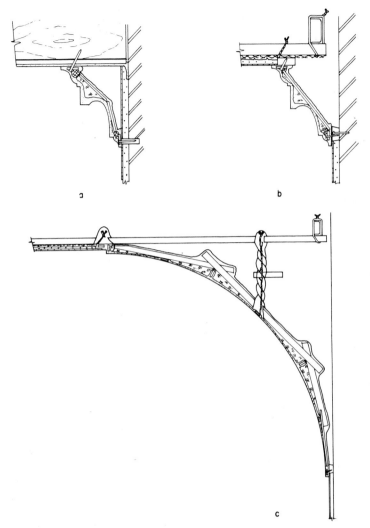

Fig. 34. Cornice (fibrous), fixing: (a) to a joist and a plug, (b) to grounds on an EML ceiling and on a brick wall, (c) a large section with rebated ceiling line to receive solid plastering

(b) fixing to a thickness batten that is fixed to the background prior to plastering, the backgrounds being brick/blockwork and expanded metal lath; (c) wire and wadding to the grid prior to the fixing of the metal lath. The section will require an extra member, the depth of the solid plastering, with a rebate to receive scrim in the finishing coat.

Cornice

The sequence for fixing is as follows:
1 The projection and depth are marked on the ceiling and wall respectively and lines snapped to them.

2 At least three blocks of wood per cast are fixed to the underneath of the wall line so that the cornice may rest on them during fixing.
3 The strike-offs are cleaned.
4 Where necessary, the cast is cut to length and to mitres.
5 The cast is lifted on to the blocks. (A cast some 4.500 m long and 0.150 m wide can be fixed single-handed, provided it is lifted on edge, wall line uppermost and face in. The wall line is placed on the blocks so that the cast is supported along its length while it is hinged up in position with outspread arms. A T-strut, nail or wire will then hold the cast for general fixing.)

When the cast is securely in position, and to line, and all the joints and mitres pick up, all nails may be punched home and all wires, mitres and joints wadded up from behind. When the wads have set, the wires are punched home and stopping may begin. For full wire and wadding techniques see PLAINFACE, fixing. For mitre cutting see MITRE.

Stopping

All rebated joints should be stopped with scrim as in PLAINFACE, stopping. The wall and ceiling-line joint to finished plasterwork should be reinforced with strings of canvas. Mitres are wadded with canvas or strings in the stopping. See also MITRE, mitring.

Solid

External

An external cornice should be functional as well as decorative. It must have a weathered or inclined uppermost surface to carry away rain water. It must also include a drip member with throating. This will help to prevent water from running down the face of the building.

In most cases an external cornice will be run using two rules; a *slipper* rule and a *nib* rule. It is also almost certain to be run on a rebate or rabbet. Metal bracketing or a purpose-built stone, brick or concrete CORE is essential as most external cornices have large projections. In the case of stone, brick or concrete a spatterdash coat is essential for good adhesion. All 'dubbing out' should be done at least one day before starting the running of the cornice.

When making the running mould for this operation, the metal profile and stock should be continued for about 25–50 mm beyond the top member and take the line of the weathering. They should then be cut vertically so as to be sufficiently large to bear on the nib rule and the face covered by a metal shoe. The rebate should be fixed so that it will give at least a 3 mm clearance from the wall surface when in contact with the slipper rule.

To obtain the position of the two rules the running mould is held in position at either end of the run and the vertical members plumbed down. Marks are made at nib and slipper positions at both ends, and lines are snapped with a chalk line. The rules, which should be 50 × 25 mm, are fixed at both ends and to the lines. Where the background or backing deviates from the straight, it is necessary to insert small packing pieces behind the slipper rule to make sure that it is perfectly straight, both on the horizontal edge and the vertical face. The nib rule should be bedded in a weak mix of cement and sand. This will have to be removed after the cornice has been run. Also, if this rule cannot be nailed in position, bricks or other weights should be laid upon it, behind its face line. This will prevent movement. A brick every 200 to 300 mm, again bedded in, will normally suffice. Struts may also be used, from the wall to the back of the rule.

After the keyed dubbing-out has hardened, a metal muffle should be fixed to the running mould, about 5 to 6 mm in front of the metal profile. The core must be run out to near perfection, using the

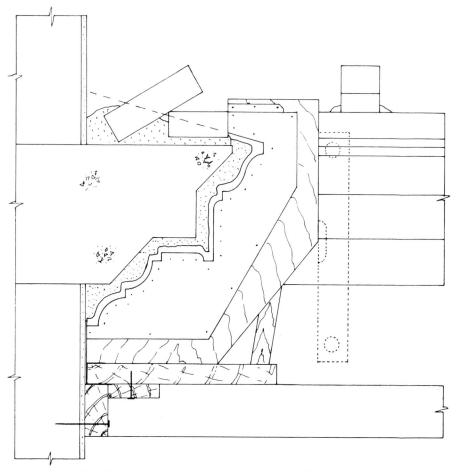

Fig. 35. Cornice (solid), external: running mould in position

wet and dry method explained in RUNNING IN SITU, Portland cement and sand. The mix for the core and the finish should be 1 part Portland cement to 2 or $2\frac{1}{2}$ parts washed sand.

Mitres and small breaks must be worked in by hand using wooden joint rules and a spare metal profile for checking the contour as the running proceeds. Larger breaks are run up using a notched or backed rule, again while the main cornice is still green.

Internal, formed – flush cove/bottle cove

A coved angle with no members, making the ceiling and wall one surface. Small coves are formed with a gauge as SKIRTING, COVED; and large with pressed SCREEDS as FLOATING TO CURVED BACKGROUNDS.

Internal, straight, run

These can be roughly divided into three categories: *small cornices*, requiring no coring out; *medium cor-*

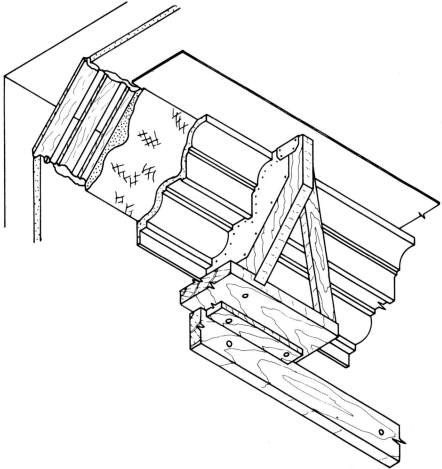

Fig. 36. Cornice (solid), internal: running on Scotch brackets

nices, requiring coring but no bracketing; and *large cornices*, requiring coring out and bracketing (Fig. 36). See BRACKETS; BRACKETING.

If brackets or coring is necessary it is usual to run the core out to a muffle. The core could consist of either sand and Class A plaster or coarse stuff and plaster. All cornices are run on a slipper rule fixed to the wall, in conjunction with either a lime putty and plaster screed or a rabbet over a floated or a finished wall. At the nib end it may be run on a nib screed of putty and plaster or the ceiling finish. The running mould is offered up in position at all angles, and chalk lines are snapped around the room. The slipper rules are fixed to these lines.

The running off should proceed as in RUNNING IN SITU, lime putty and plaster. It is generally thought that alternate sides of the room should be run off together but many craftsmen prefer to run a room off in a continuous system. Once the operation on a length has begun, it should be completed.

Delays can cause furring when running off and peeling when the moulding is finished. (Gauges should be consistent and well mixed, applied evenly to the entire length and girth of the cornice, with special emphasis to the top and projecting members.)

On completion of the running, all droppings and selvedge must be cleaned off and the mitres cleaned out. All nails should be removed from the rules and these should be laid flat so that they will keep straight.

If a rabbet is required a 2 to 3 mm gap is necessary between the slipper and wall face. The running mould is horsed up so that the stock is central to the slipper and well braced. Metal shoes are necessary at the slipper ends and the nib.

Also see under specific moulding to be run.

Internal, curved, run

The running off of a cornice to a curved wall will be as for straight with the following exceptions:

1   Pliable running rules are necessary so that they will follow the wall line accurately; no screeds are necessary.
2   The slipper should be shaped so that it will follow the concave portion of the wall, and two pegs inserted in the bottom of the slipper – one at each end to act in a similar way to a rabbet.
3   The horse and profile mounted at the leading edge of the slipper, though this is not always considered necessary.

(See Fig. 37.)

Circular curve. Stock may be mounted in the centre of the slipper.

Changing curves. Stock must be mounted at the front of the slipper in order that the stock remains in contact with the wall.

CORNICE DENTIL-BLOCK

A cornice containing a row of DENTIL BLOCKS as a line of enrichment. The cornice must be cast from a mould having the line of blocks reproduced by a pliable moulding material. The types of mould that may be employed are clay case, run case and insertion. See MOULD, REVERSE.

The cornice is cut to mitre with a block on the corner at external mitres and either with one in the corner of internal mitres or with two butting. If the enrichment is not cast strictly to size, the blocks are left off the end, either by masking them off in the mould or by cutting them off the cast and making good the bed. The bed is then completed with the rest of the mitre, and the blocks bedded in to balance correctly by slightly increasing or decreasing the gaps between them.

CORNICE, MODILLION-BLOCK

A cornice containing a row of MODILLION BLOCKS as a line of enrichment. The blocks are planted on a soffit with their projection exceeding their depth and spaced so that a square is formed between them. The particular problem with this type of enrichment is to balance the blocks in the mitres. With all but the smallest of blocks an insertion or run-case mould (see MOULD, REVERSE), which allows stretching and shrinking, is impracticable. Cold pour is too expensive and if PVC is used the excessive thickness of pliable material between the blocks will cause distortion on the face when it shrinks on cooling and maturing.

The mould most commonly used is therefore a clay-case mould, and blocks are left off the end of the cast so that they may be re-spaced to balance after the cast has been fixed. Blocks are blanked out by filling them solid with plaster and ruling it in to a smooth finish, flush with the bed. They are then shellacked and greased, and will form a continuous bed in the cast on which subsequently cast blocks are fixed by filling the hollow block with part of a WAD and pushing the rest through a hole cut in the soffit. The block is held or strutted in position until the plaster picks up.

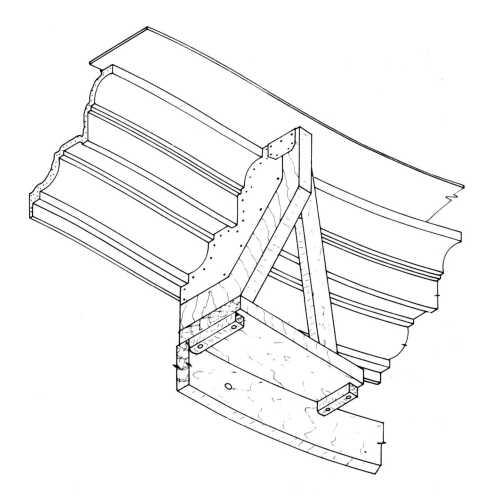

Fig. 37. Cornice (solid), internal: running on curved walls

COTTLE

Old term for the wall round an open mould.

COUNTER LATHING

Where timber faces of items such as wood joists, beam soffits and wooden lintels exceed 75 mm, and wood or metal lathing has to be fixed to them, counter lathing should be applied. Pieces of 6 mm wood lath or batten should be nailed along the entire length of the wooden member. This will provide a gap behind lathing subsequently applied, allowing plaster to pass through to form a key. See also BRANDERING.

COVE CORNICE

A CORNICE that is CAVETTO in section.

COVE, FLUSH/BOTTLE COVE

See CORNICE, solid, internal, formed.

COVE, GRANOLITHIC

See SKIRTING.

COVE, REFLECTOR

See LIGHTING TROUGH.

CRACK

See FAULTS.

CRADLE

A wooden casing for steel joists and girders.

CRADLE, LEVELLING

Used to level in ceiling surfaces divided into separate bays by BEAMS. (See Fig. 51 (c).) See FLOATING TO BEAMS.

CRAZING

See FAULTS.

CROSS-BRACKET

In fibrous plaster casts. See BRACKETS; CORNICE.

CROWN

The highest point of the mouldings forming an arch.

CUPOLA

A hemispherical domed ceiling or roof placed over square or multi-angular buildings.

CURING

The hardening of a material over a period of time.

Cement

The curing is accomplished by keeping the mix damp, either by applying water by soaking or spraying, or by preventing evaporation by covering with polythene, etc

Polyester resin and cold-pour moulding compound

Curing agents are added to make the material harden. See under specific item.

CYMA RECTA AND REVERSA

See MOULDING, SECTIONS.

DADO

The portion of a pedestal between the capping and the plinth. Also the part of a wall surface below a dado moulding and above the skirting. Usual height of dado moulding is 1 to 1.075 m.

DADO MOULDING OR FLUSH BEAD

A dado moulding may be in the form of a flush bead (see •RAMP, Fig. 95) or a projecting panel-type moulding. In general they are run in putty and plaster but can be run in Keene's cement. Some time ago, all mouldings were run on two putty and plaster running screeds, but this is not always the case today. However, with dado or flush bead it will largely depend upon the finishing materials used above and below the moulding line.

If the setting above the dado is sufficiently hard to resist nib wear then there is no reason why one should not run on this. But two rules must be observed. Firstly, the finishing coat must be ruled in horizontally to perfection, otherwise every wave and dip will transmit itself to the moulding. Secondly, the running mould must be made in such a way as to ensure that the top member is on the same line as the setting coat: if it is 1 mm behind there is a nasty edge to make up; if 1 mm in front, a good straight run arris has to be cleaned off.

The slipper should always be run on a rabbet. This will make screeding unnecessary but the rule must be perfectly straight both horizontally and vertically.

Whether run in putty and plaster or Keene's, it must be cored out first. In the case of a flush bead the lower member will act as a guide or screed for the setting. A projecting moulding, when run on a rabbet should automatically form a setting member.

DAMP, EFFECT ON PLASTER

Internally, all lime and gypsum plasters will eventually disintegrate under conditions of perpetual dampness. Occasional damp will cause staining on the surface. Application of a damp-resisting liquid to the damp areas will only accelerate the breakdown by confining the dampness. Such areas should be hacked off and re-plastered in a waterproofing mix.

Reasons for the damp should be discovered and eliminated if possible. Externally, continual dampness will have a similar effect on certain cement–lime–sand mixes but transitory dampness, e.g. from rain, should not result in serious harm. Dampness, particularly bypassing a damp-proof course on the ground, can bring with it salts that cause EFFLORESCENCE.

DASHING, WET OR DRY

See FINISH/RENDERING, EXTERNAL.

DATUM LINE

A struck level line used to ensure that all finished plastering is level and square. It is usually struck around a room or building to marks obtained using a water level. The position of the line is at a convenient height for working to, and all setting out is done from this using datum rods. These rods are occasionally fixed to squares, to assist in squaring in adjacent surfaces in the same plane.

DATUM ROD

A piece of 50×25 mm timber used to measure distances accurately from the datum line when setting out in both solid and fibrous plastering. One end is cut square, and accurate measurements of various features are made from this end which is then held flush to the line; the position of the marks on the rod are then transferred to the wall surface. For large jobs where there is a great deal of setting out, the marks should be saw cuts and plainly identified as to what feature they represent. Alternatively several rods may be used, but these too must be plainly marked otherwise mistakes can occur by using the wrong rod.

Some craftsmen prefer not to have the datum-line mark at the very end of the rod. They use a saw cut, well in from the end and again well marked for identification. An advantage in having the datum mark on the end is that should one wish to trace a curved line, a nail on the line will act as a pivot for the rod.

DAUB

See WATTLE AND DAUB.

DENTIL BLOCK

A rectangular block forming part of a row of such blocks giving a line of enrichment in a dentil-block cornice, mantelpiece, overdoor, pediment, etc. It originated in the classical orders of architecture. The width of the block is two-thirds of the depth, and the space between the blocks one-third of the depth. The projection of the block is the same measurement as the width.

Production

(i) A length of moulding is run to the section of the blocks. The moulding is then cut square across its width into the block sizes, allowing a little extra for rubbing down. The sides of the block are rubbed smooth on a sheet of glass or a flat marble slab, using water as a lubricant and checking for square. The blocks are then stuck into their bed, checking that they are square to the model and using a gauge between them for uniform spacing.

(ii) The moulding from which the blocks are to be cut may be run on their bed as a run loose-piece and the procedure is then as for (i) above.

Moulded dentil blocks

Dentil blocks may also take the form of small, plain MODILLION BLOCKS but, unlike the modillion, do not alternate with PATERAS or have the small moulding breaking round them.

As the dentil-block enrichment must be reproduced by a pliable moulding material, it must first be laid down as a model. For reproduction see under specific feature containing the blocks. For raking dentil block see RAKING.

DENTIL-BLOCK CORNICE

See CORNICE, DENTIL-BLOCK.

DEPETER

Pebble dashing (see FINISH/RENDERING, EXTERNAL) pressed in by a wooden float while still green. Can also be sea shells, coloured glass fragments, etc.

DEXTRENE

Made into a solution and added to the firstings of plaster casts. It hardens the face of plaster of Paris and incidentally retards the set slightly.

DINGING

A cheap, one-coat finish in Portland cement gauged coarse stuff – 5 to 6 mm of this material is laid on to clean, dampened brickwork and finished with vertical stock brush strokes.

DOME, FIBROUS

Fibrous plaster domes are cast from moulds that have been spun (see SPINNING) or formed by RIBS as a DRUM. Both its geometrical shape (plan and cross-section) and its size will dictate the method by which its reverse mould is made.

Circular on plan

Reverse moulds for such domes (Fig. 38), whether semi-circular, segmental or semi-elliptical in section, are best spun from a pivot block whenever possible. The size of the casts will determine the portion of the dome that is to be made up in reverse, a sector of a 3 m radius dome being quite manageable in one. The internal shape of dome moulds dictates that the muffled cores of all but the smallest domes must in turn be cored out for running with lathed-out ribs. Even if the size of the section allows only a rough, woolly finish to be obtained, it nevertheless provides a perfect line. The surface is brought to a good finish by busking in creamy plaster and this is the normal way of finishing a drum ruled in from ribs.

Sectors of larger domes (Fig. 39) may be run up in separate sections which, if they are kept to a metre or so high, may comfortably be run about 2.5 m long to a radius limited only by the shop facilities. With domes of semi-circular section the same profile may be used to spin each portion of the sector, by merely altering its attitude to the horizontal by offering it on to a full-sized section of the dome. With a semi-elliptical section, however, each part of the section – being different – will have to be cut and filed afresh.

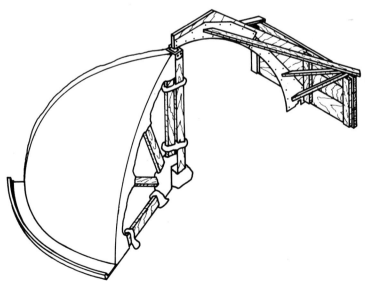

Fig. 38. Dome (fibrous): reverse mould for a sector of a dome circular on plan

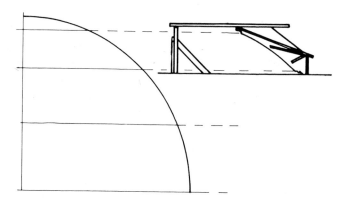

Fig. 39. Dome (fibrous): running mould to spin a part of a sector for a dome circular on plan

On domes semi-circular in section that are too large for each part of them to be spun from the true centre (Fig. 40), all curved plainface needed to form any part of the dome may be cast from the same piece of spun reverse mould. The concave shape of a dome is constant for any given position on the dome. Therefore, if the shape of the plainface is set out, all pieces may be cast from the same portion of the reverse mould.

Having spun a reverse mould with a profile cut to the correct curve, the curved rules that will form the horizontal joints between the plainfaces are determined by spinning a plaster rule to the horizontal radius of each joint. These rules should be prevented from twisting in any direction by wadding on timber braces. When the rules are placed on the reverse mould, they will automatically lie over the face in their correct positions. They must be

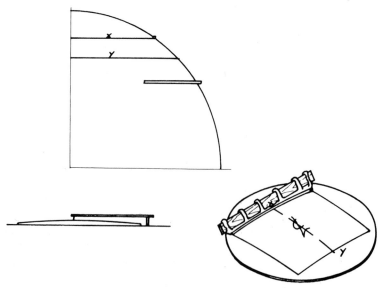

Fig. 40. Dome (fibrous): the cap of a dome to cast all sections

positioned square to a centre line, and the length of each curved rule calculated and marked centrally on each side of the centre line. The straight end rules may now be fixed between these and, although technically it should also be curved, the curve is so slight on a dome requiring this method of construction that the widened gap can easily be taken care of in the stopping. As the rectangular section of the spun rules will not be in the correct plane, they are used to set out the casting lines only and casting rules are fixed to these, preferably to provide lapped and rebated joints.

of the ellipse as the profile on the running mould. The dome may be cast in four quarters from the same piece of reverse mould by merely interchanging the perimeter rule with the rule producing the lapped and rebated joint for the centre.

Larger domes (Fig. 42) may have the reverse

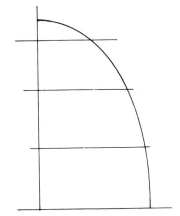

Fig. 42. Dome (fibrous): to spin a part of a sector for a dome elliptical on plan

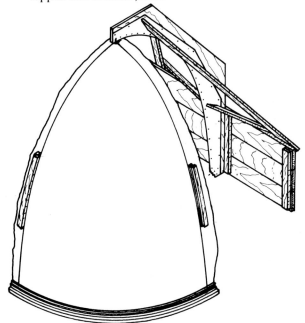

Fig. 41. Dome (fibrous): reverse mould for a quarter of a dome elliptical on plan

Elliptical on plan (semi-elliptical major axis, semi-circular minor axis in section)

Many methods exist to produce such domes (Fig. 41). A dome sufficiently small (up to some 4 m major axis) may be spun from a pivot block using a quarter

mould spun to each portion of a sector as in Fig. 39. The only difference is that the profile, being elliptical, will have to be cut and filed afresh each time.

Figure 43 illustrates an alternative method to the previous one, forming the reverse mould from ribs as a drum. The ribs are set up against a band spun to a true elliptical curve with the aid of a trammel. The band may contain a small portion of the dome, the ribs being suitably rebated to receive it, or it may merely provide the strike-offs for the reverse mould, the ribs being butted against it. It is good policy to cast a rib against the band so that it can be stood up to form a perfect line to the top of the reverse mould. This will allow all four quarters of the dome to be cast by interchanging the perimeter rule with the lapped and rebated top rule.

Elliptical on plan (semi-elliptical major and minor axes in section)

In Fig. 44 we see an adaptation of the previous method – a perimeter rule spun with a trammel – providing a good line and ribs of the required shape erected off which the DRUM is ruled. With this form of dome, however, a full half is needed. A cast from one quarter of the dome will only fit in the identical spot in the corresponding diagonal quarter.

Casting

The dome, whether cast whole or in sections, is cast in the same way as curved plainface. The fixing operation is greatly assisted if provision is made in the mould, or with casting rules, to provide sections of domes with the appropriate lap-joint arrangement. Jointing with surrounding plasterwork should also be provided for, with rebates round the perimeter of the dome. Casts to make up domes are plainface subtly curved in all directions and therefore should be prevented from twisting by suitably wadding on 25×50 mm battens which can be made to serve as fixings, the cast being hung from them off fewer but stronger fixing bearers.

Fixing

Small domes made self-supporting by integral stiffening battens may well be fixed in timber ceilings by screws and nails, but normally a dome, being the shape it is, is supported by hanging it by its battens from metal fixings running across and above the dome. They are therefore reinforced accordingly and fixed by wire and wads, with all joints between casts wadded up as barrel CEILINGS.

When a portion of ceiling some 100 to 200 mm wide is cast as a part of the dome, the band will hold

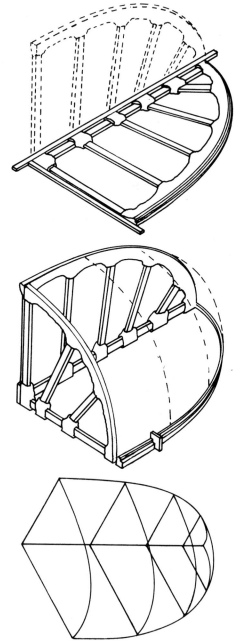

Figs. 43 and 44. Dome (fibrous): reverse moulds for elliptical domes formed from ribs as a drum

the rim rigidly to a true line. With very large domes, however, it is a great help to have a radius rod fixed to a pivot or trammel at the centre of the dome from which the rim is regulated. Very often the rim of a large dome cannot be boned as a whole but any kicks at the joints between casts on the rim will stick out like a sore thumb. If casts cannot be made to pick up properly due to distortion, etc., a sweet line may be obtained by tacking up or strutting from the scaffold a rule round portions of the curve at a time, planing off any surplus plaster and well keying and filling out any hollows to the rule.

Stopping

All joints between casts should have rebates which are well keyed to receive plaster-soaked scrim and are stopped and made good as for PLAINFACE, using either a curved joint rule or busk to rule in the joints. With domes circular on plan, it is well worthwhile filing a 300 mm curved joint rule, even if it is merely cut from stiff busk, as it can be used on edge to give a true surface in any direction.

DOME, SOLID

The choice of method will normally be dictated by the overall size and shape of the dome. On the largest variety, the use of dots and pressed screeds as in FLOATING TO CURVED BACKGROUNDS is the most practical method. For medium domes a run lip or rim around the base of the dome is usually practicable, while for the smaller types an entirely run-up section may be possible (Fig. 45). Accurate setting out is essential. The true centre point must be located and, no matter what shape or size the finished item is to take, all measurements must proceed from this.

On large and medium domes, it is usual to fix a vertical bearer from the centre point so that it drops to at least the same level as that of the ceiling surrounding the dome. This must be well fixed at the top and securely strutted along its length. For work on large domes, dots can be gauged in from this bearer and screeds pressed to these dots using a curved floating rule cut accurately to shape. At other times it is cut off, level with the surrounding ceiling, and a centre is fixed into the base end. A

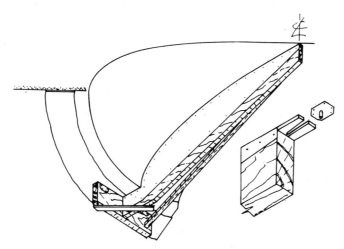

Fig. 45. Dome (solid): spinning a full dome

mould is constructed to run a lip or rim around the complete circumference of the dome, and a gigstick is fixed to the mould to bear on the centre bearer.

So that the mould will run smoothly the surrounding ceiling must be floated to a true surface, lined in to a DATUM LINE and accurately levelled to the correct measurement from the apex of the dome.

pletely. A running mould is made to the true section of half of the dome. It is cut so that one end will bear on the flat ceiling and the other end will pivot on the centre point. This can be either a firmly fixed centre at the apex of the dome or, as before, on the base edge of the central bearer. A muffle or lath is fixed to the profile and the floating run up to this.

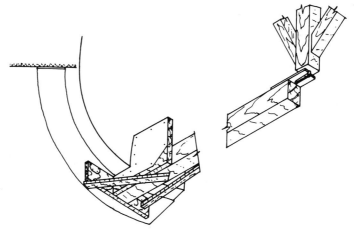

Fig. 46. Dome (solid): spinning the rim

The running mould can be constructed by setting out a good external angle, flat on the ceiling side and curved on the dome side. This should be some 50 to 75 mm either side of the arris. It is then cut back on both sides to form setting members.

On the dome side, the profile can then be cut to continue the curve for another 50 to 75 mm so that the run section will form not only a true circumference and good arris, but also a screed for the rest of the dome to be floated and set to. The running mould will bear only on the flat ceiling at the slipper point, and the central bearer at the gigstick pivot point. Therefore a lime putty and plaster screed will be necessary on the ceiling around the perimeter of the dome. Floating proceeds as before, once this rim has been run (Fig. 46).

On smaller domes and even some fairly large ones that are shallow the floating can be run up com-

The muffle can then be removed and the finish coat run up. In most cases it will be possible to get a perfect finish to the dome surface by running and an old well-worn finishing trowel can either be cut to fit, or the central rivets can be deliberately sprung so that when pressure is applied the trowel takes the shape of the curve.

All dome shapes other than circular in plan must be formed in the dot and pressed screed method. These will require accurately cut curved rules, possibly to several shapes; therefore they should be numbered or lettered so that the right rule will be used for the right area. Dots are gauged in around the circumference spaced to suit the curved rules. Then dots may be placed between these circumference guides and the main central or apex dot, so as to form sectors.

Pressed screeds and general floating to curved

surfaces can proceed from this stage. The finishing coat is applied when the floating has set, ruled off with the curved rules until a perfect shape has been obtained then laid down and trowelled with the adjusted trowel, once it has hardened. It may be necessary to go over the entire surface with a piece of suitable busk.

Where mouldings, or even a small cornice, appear around the circumference of the dome, they should be run up in the same way as one would run up the rim or lip. The gigstick would have to be strengthened to resist the pressure when being run, and of course on large domes no attempt should be made to run off in one. Either core out well to a muffle and then finish or run in shorter lengths, say a quarter or half the circumference at one time.

DOTS

A mound of plaster applied to the background or backing in an appropriate position, to assist in obtaining a true surface in plaster. It may have a small piece of wood lath or something similar embedded across its face, or it may be a nail, or several nails, driven into the background. It is used as a ground for the level, plumb rule or square.

Plumbing down to walls

When dots are to be used for plumbing down a wall they must be positioned correctly. The best possible position is 150 mm away from all internal angles – wall, ceiling and floor. One dot is put at this distance below the ceiling and away from the vertical angle, while another is put at the same measurement up from the floor and away from the vertical angle. The two dots should now be directly in line, one above the other. Two methods are used to check their being plumb: in normal-sized dwellings, where the dots are probably a little over 2 m apart, a parallel rule is held at right angles to the background, so that one face presses on to the two dot faces while the parallel edge is furthest away from the dots. A spirit level in an upright position is held on to this. If the vertical bubble does not line up perfectly the offending dot is either pressed in or built out until the bubble shows perfection.

Another system is used on larger buildings where the two dots may be many metres apart vertically. Two timber gauges are made as illustrated in Fig. 56(c). It is essential that the distances from the narrow ends to the shoulders are identical for the two gauges. A long, pre-stretched line with a heavy plumb bob at one end is tied at the wide end of one gauge. After the two dots have been placed in position, the length of line is measured so that when it is held on the top dot the plumb bob will hang free, just below the bottom dot but above the floor. The uppermost gauge is then held in position so that the narrow end of it is on the top dot, central to its length and at right angles to it, with the line falling down accurately in the shoulder. As the bob swings free at the bottom, the other gauge is held in an identical position on the lower dot; as soon as the line drops in line with this gauge's shoulder both dots are in a plumb position. It may be necessary to press one dot back or build one out until the required position is obtained.

In both cases the procedure is repeated at each end of all the walls that are to be FLOATED. Dots can be placed horizontally with vertical screeds, or vertically with horizontal screeds. In either case, the original positions for the dots will remain as already described, and the plumbing down will be as before. There is no definite ruling on the choice of horizontal or vertical dots: craft experience is the best guide. However, it is generally felt in the trade that horizontal dots and vertical screeds are more accurate for larger operations, while vertical dots and horizontal screeds are suitable for smaller jobs.

DOTS, DRY LINING

Most BACKGROUNDS that are to be dry-lined will need levelling, and one method of obtaining this is to apply a purpose-made dot at regular intervals to receive the plasterboard. These dots are bitumen-impregnated fibre-board pads and are held in position by Class B board finish plaster or lightweight bonding plaster. These dots are lined in at 450 mm centres along the wall 230 mm down from the ceiling and 100 mm up from the floor, with a dot midway between the top and bottom dots. More than one dot must be used where the space between their centres would be greater than 1070 mm. See FLOATING, Fig. 56 (b).

DOTS, LINING IN INTERMEDIATE

Walls longer than the length of the floating rule will require dots to be placed between the two at either end. These are put into a line stretched across the face of the two dots. The line is held in place by a nail outside each dot. Three pieces of lath are used: one piece wedged between the first dot and the line, one between the last dot and the line, and the third as a feeler to regulate the intermediate dots.

Dots so positioned will have either horizontal or vertical screeds wiped in between them with the laths positioned accordingly. See FLOATING, Fig. 56 (a).

DOTS, LEVELLING IN

Intermediate dots required with horizontal areas are levelled in from the perimeter dots with a floating rule and spirit level (Fig. 57).

DOTS, SQUARING IN

The method of placing dots level on a horizontal surface by striking them from a datum line marked on adjacent vertical surfaces (Fig. 57). To level a ceiling, a dot is positioned at the lowest point on the ceiling to a square held on the floated wall. The position of the DATUM LINE is marked on the square, and all perimeter dots round the ceiling regulated by the square held on the line. These dots are also square to the surface off which they are struck. In the same way, a floor may be levelled in or sloped to falls by gauging from a datum line snapped near the floor. Dots the full depth of a beam may be squared down its sides from a floated ceiling (Fig. 51 (b)).

DOUBLE-HORSED

A running mould that has no nib or nib slipper but is horsed so that the function is symmetrical about the stock, e.g. the slipper on each end of the stock will run against a running rule as well as on the running surface. See also RUNNING, DIMINISHED.

DOWEL

The male section of joggle and dowel. Usually a circular piece of reinforcing rod set into locating holes in items to be fixed together, e.g. baluster and balustrade.

DRAUGHT

The splay given to a vertical member so that it will draw from a mould. If a splay is unacceptable it may be planed off or cut away with a purpose-made gauge.

DRAWINGS, WORKING

See PRODUCTION MANAGEMENT.

DRESSING

See ENRICHMENT, planting.

DRIERS

See RUNNING IN SITU, Portland cement and sand.

DRIP

See MOULDING, SECTIONS.

DROP BOARD

A clean scaffold board or the like, laid on the floor at the base of the wall to be plastered. Its use is to catch the plaster droppings, so protecting the floor and enabling the operative to pick up any dropped plaster and return it to the spot board.

DRUM

A drum is a curved surface formed by ruling in from RIBS. It may be used either as a reverse mould or a support for a reverse mould, such as a curved BEAM CASE or pliable flood mould, or it may be a model from which a reverse mould is to be cast.

Basic construction

First the required number of RIBS is cast to the correct shape and containing all the rebates required to receive the laths during the lathing out of the drum after the ribs have been fixed in position. The drum is set out on the bench with the position for each rib made ready by fixing blocks and cleats. Ribs are then positioned against the blocks and temporarily fixed square or plumb. The ribs are then wadded to the bench and securely fixed together, with 25×50 mm braces. The top surfaces of the ribs are shellacked and greased and the areas between them lathed out. Canvas is cut, soaked in neat plaster and laid over the laths.

When set, the areas between the ribs can be filled with plaster. This is done by ruling in with a gauge rule, striking down 3 to 6 mm below the surface. The surface is well keyed and allowed to expand fully before finishing when it is brushed to remove all loose particles. The finishing plaster is gauged to a creamy consistency and rubbed or brushed immediately into the key. For ruling in, forming and shellacking such a surface see BENCH, construction of plaster benches.

Points to watch

Expansion of the plaster between the ribs is perhaps the most troublesome aspect of drum construction as this causes swelling and a uniform bulge is produced. To counteract this the following steps may be taken:

1　The surface is formed in two layers, the bulk of the expansion relieving itself in the floating with the topping layer being kept to a minimum.

2　Distance between the ribs is kept down to about 600 mm.

3　As the topping picks up, a lath is drawn over the drum through its entire thickness, midway between the ribs, in an attempt to allow the expansion to relieve itself sideways, closing the groove slightly rather than bulging upwards.

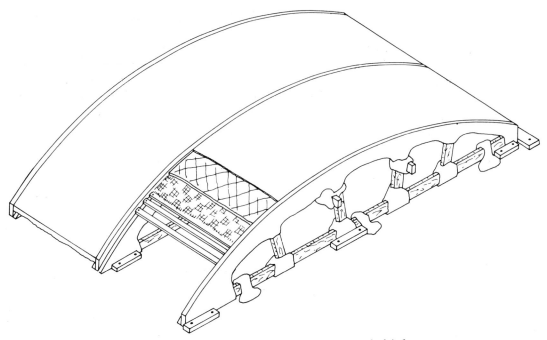

Fig. 47. Drum (fibrous): segmental for casting curved plainface

4   When ruling in the finish a feather edge with a slight bow may be used instead of a straight edge. The resulting hollow surface will become straight on swelling. Experience will show the degree of bow required. This could be 1 to 2 mm in 600 mm, depending on the thickness of the finishing coat, stiffness of the plaster and the degree of suction afforded by the floating coat. *To obtain a perfectly flat drum* it will be necessary to scrape the drum, preferably before the surface is joint-ruled in. This is done by colouring the edge of a wide straight edge and scraping it over the drum. Everywhere the rule touches it will leave colour. These areas are then scraped off and the operation repeated until the whole surface shows colour. For use see BARREL; BEAM CASE; LUNETTE.

Double-curvature drum

This type of drum requires either a curved rule in the surface between the ribs or a pliable rule that may be bent over the undulating ribs, thus splining in the line. To bend a rule to the best possible line it should, wherever possible, traverse four ribs – one each side of the pair being filled in. For use see DOME.

DRY LINING

For preparation see DOTS, DRY LINING. When these have set hard it should be possible to apply the plasterboards.

Pre-mixed lightweight aggregate bonding plaster or Class B board finish is placed in a series of dabs with the laying trowel between all dots vertically. Each dab should be about trowel length and proud of the dot thickness. Both width and gap between dabs should be about 50 mm. Apply only sufficient dabs for one board at a time and make certain that both dabs and dots at the joint between two boards will bridge the gap equally so that each board will get the same fixing.

The board, cut 25 mm short of the floor to ceiling height, is then offered up in position. A foot lift, which is a piece of triangular-shaped wood using the seesaw principle, will assist with this. Place the board in position with the foot lift beneath its bottom edge. Apply pressure with the foot on the lift and push the board up tight to the ceiling. Then tap back firmly till the board is in contact with every dot. The dabs will be flattened against the board and good adhesion should be obtained.

Check that the leading edge is central to the dots and plumb; and then nail in position. The best nails for this purpose are the double-headed nails made for the job. Drive them in to the first head; when the plaster has set, the nails can be removed with pincers and used again and again.

Plaster dabs are applied for the second board and the operation continues till the area is covered. All joints should butt and this includes angles. Where cutting is necessary the cut edge should be hidden in the angle. Nailing will only be necessary at dot positions on the board edges.

For all narrow widths, cut the boards for these so that they project in front of the adjacent wall face. Bed them up with plaster, ensuring that they are plumb, square and marginable. When these have set apply dots to the wall face so that the boards will finish in front of the leading edge of the reveal board. At all times it is essential to have a bound edge one side of an external angle, this can then mask the cut edge. Where cut edges are unavoidable, sand paper lightly to remove any rough edges and leave a 2 to 3 mm gap.

Narrow widths up to 460 mm need no dots or pads. The jointing to dry lining is normally covered in one of three ways. For the taper-edged wallboards a flush surface is usually obtained by taping and filling flush. The square-edged boards are usu-

ally butted, and the joint covered by a purpose-made strip. In the case of the bevel-edged boards, the bevel on two boards meeting becomes a V and this may be loosely described as a decorative feature.

To joint the taper-edged boards, apply a continuous band of joint filler to the depression caused by the two taper edges. Make certain that where small gaps occur, filler is pushed well into these, also that there are no areas left uncovered. A purpose-made applicator is best for this job. Then press the joint tape into this, using taping knife. Ensure it spans the joint and is firmly embedded, then apply another coat of filler over this, bringing it flush with the board surfaces. Clean off any surplus material with a damp clean sponge.

After about an hour apply a thin coat of joint finish to a width of approximately 200 mm, and feather the edges out with the jointing sponge. When this has hardened repeat the operation and finish with the sponge as before. All angles, internal and external, must be taped in the same way, also cut edges and other types of board.

See also BACKGROUND, PLASTERBOARD.

DRY-OUT

See FAULTS.

DUBBING OUT

The application of material over and above the normal thickness of a backing coat. It is normally found where a background is badly out of true. The background should be thoroughly cleaned, keyed and covered with a sand – cement – lime mix. This, in turn, must be well keyed to receive the backing coat. Where the hollows are slight they can be filled with the backing mix, well keyed and left to set before being floated.

Dubbing out is also necessary in forming raised or sunken panelling in solid, on the panel area for raised panels and on the wall surface between the panels for sunken work. In each case the areas to be dubbed must be well keyed with a comb scratcher and left till set.

Dubbing is necessary, too, at internal angles to a waterproof tank or room.

See WATERPROOFING, internal.

DUST See Appendix 1, p.279.

DUSTING

1 In casting – see CASTING IN FIBROUS PLASTER, one-gauge method.
2 Fault in cement floors – see FLOOR LAYING AND ROOF SCREEDING.

EAVES, PLASTERING

The eaves of a building project beyond the face of the wall at the lowest part of a pitched roof.

Because of their position and construction these normally have a metal lathing background. They should be plastered in two coats, render and float. Both mixes should be 1 : 1 : 6, cement : lime : sand, and should contain a quantity of hair, sisal or similar fibrous material. The rendering should be applied to a thickness of 7 to 9 mm from the face of the lathing and be well pushed through. It must be well scratched and left for a minimum of 24 hours. The second coat should be of the same thickness, ruled straight and finished with a wooden float as in cement PLAINFACE.

ECCENTRIC RULE

See RULE.

EDGING BEAD, PLASTERBOARD See BEAD

EFFLORESCENCE See FAULTS.

ELECTRICAL SAFETY See Appendix 1, p.282.

ELLIPSE

Running with a trammel

A rod or gigstick can be made to trace the path of a true ellipse by sliding two points on the rod along the major and minor axes. The two points are determined as shown in Fig. 49 (a). A running mould, fitted with its profile at the end of the rod, will run an elliptical moulding.

Limitations of this method

When the rod is at the 'corners' of the ellipse (Fig. 48), it is not at a normal to the curve and the resultant twisting of the profile will reduce the width of the moulding. The degree of noticeable distortion depends on the ratio between the width of the moulding and the size of the ellipse. Where only one face of the curve is required – as with a plaster rule off which a RIB is to be cast – this fault will not matter. *When running a whole, shallow ellipse* the nib of the running mould may foul the trammel at the end of the minor axis.

When making up the trammel (a) choose a substantial batten (25 × 50 mm) from which to cut the blocks; (b) fix the rules to the batten to ensure a tight fit and continuity across the intersection; (c) use rules the same depth as the blocks to prevent the latter twisting sideways under lateral pressure; (d) cut the blocks long (150 mm or so), providing

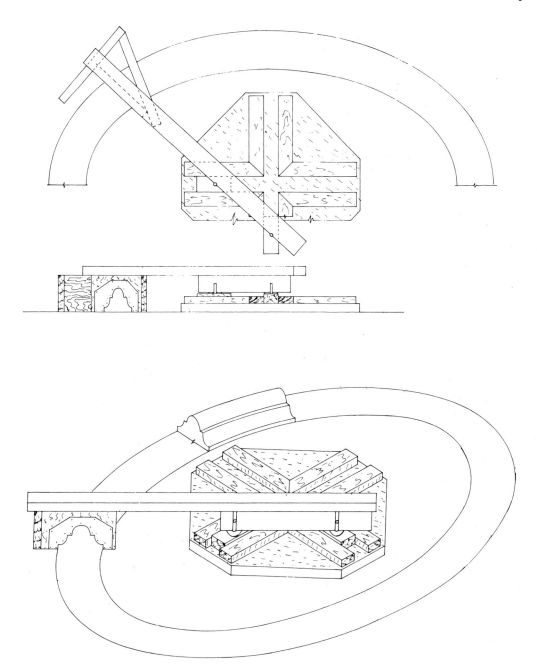

Fig. 48. Ellipse: spinning with a trammel

they will not foul each other at the intersection – short blocks will tend to twist and jam; (e) the arrangement may be made to work more easily if the trammel is constructed on a board (plywood, blockboard, etc.) rather than on the running surface, in order that it may be lifted to a convenient height to suit a conventional gigstick fixed to the top of the running mould.

For running with an eccentric rule see RULE, ECCENTRIC. For running with peg moulds see RUNNING CHANGING CURVES.

Setting out

Both half the major and half the minor axes are marked on a rod from the point that is to trace the curve, as in Fig. 49(a). The rule is traversed along the major and minor axes, keeping the point made by the minor axis on the major axis and the point made by the major axis on the minor axis. This is the most useful method of setting out a true ellipse as there are no setting-out lines to confuse a later use.

Figure 49(b) shows the 'plasterer's oval'. This may only be set out in a rectangle of the proportion three long to two deep. The centre point for the curve forming the end is where the cross in one-half of the rectangle cuts the major axis. The larger curve is spun from the end of the minor axis. The curves intersect on the diagonal drawn through their centres.

An approximate ellipse constructed to any

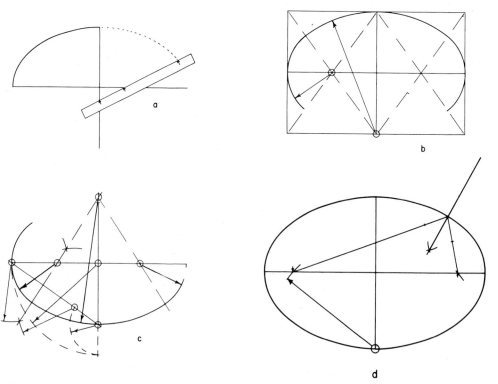

Fig. 49. Ellipse (geometry): (a) setting out with a trammel, (b) the 'plasterer's oval', (c) to spin with compasses, (d) to plot a normal

combination of major and minor axis is shown in Fig. 49(c). Half the major axis is spun from the centre on to the produced minor axis, thus showing the difference in length between them. This distance is spun from the end of the minor axis on to the line drawn from the end of the minor axis to the end of the major axis. The remaining portion of the line is then bisected. The bisector is produced so that the point where it cuts the major axis is the centre for the end curve and the point where it cuts the produced minor axis is the centre for the large curve. The curves intersect on the bisector, this being drawn through their centres.

Focal points to an ellipse are plotted, in Fig. 49(d), by swinging half the major axis from the end of the minor axis to cut the major axis.

A normal may be drawn to any point on the curve by bisecting the angle made there by the two focal points – Fig. 49(d).

ENGLISH COTTAGE

See FINISH/RENDERING, EXTERNAL.

ENRICHMENT

Lines

Anthemion: line of honeysuckle blossom.
Bead: a continuous row of beads, smooth round or carved, on an astragal section.
Bead and reel: three beads alternating with a reel (a bead with the same width but elongated), smooth round or carved on an astragal section.
Dentils: see DENTIL BLOCK.
Diaper: applied to a line of geometrical patterns.
Egg and dart/leaf/tongue: the bottom half of an egg in a cup, alternating with a dart, leaf or tongue,

carved on an ovolo section. There are many variations, e.g. the egg may be carved as a scallop shell.
Fret: straight lines forming a geometrical pattern. Usually appears the same when inverted.
Guilloche: interlacing circles.
Leaf: (i) Waterleaf – continuous, one or two leaves alternating, carved in low relief on a cyma reversa section. (ii) Acanthus – continuous, one or two leaves alternating, carved in low relief on a coved section. (iii) Feather – continuous, one or two long, feather-like leaves alternating, carved in low relief on a coved section.
Modillions: see MODILLION BLOCK.

See also MOULDING, SECTIONS.

Planting

Lines of enrichment may be of COMPOSITION or be cast in plaster from small, pliable flood moulds. DENTIL BLOCKS are usually cut from a run section. The enrichment is usually cast some 200 to 300 mm long to repeats. Lines of enrichment on a model are so balanced that they will pick up with each other and balance when an internal or external mitre is cut at the ends.

If PVC or gelatine is used, a precision back on the enrichment will produce a join between it and its bed that will not require stopping. With hot-melt compounds no air pocket, e.g. cavities under enrichment, should be present. The air will expand on heating and force its way to the face of the mould, probably via joints and stopping. This will lift the skin while the bulk of the material is still soft, causing a depression in the face of the mould. A slight nick on the profile will give a line on the model to which the enrichment can be bedded when no member is present to plant it to.

ENTABLATURE

In classical orders the main beam carried by the

columns. Moulded to form the cornice, frieze and architrave.

ENTASIS

The tapering, with a convex curve, of the shaft of a COLUMN or PILASTER. It may be from cap to base or limited to the top two-thirds.

The left-hand side of Fig. 50(a) shows the height set out with the base radius RB and the top radius RT. RB is scribed from P to cut the centre line at C. A line is then drawn through CP and produced to cut the base line of the entasis at B. Any number of lines may then be drawn from B to cut the centre line. RB is then marked along each line from the centre line, thus causing the shaft to narrow towards the top as the angle increases.

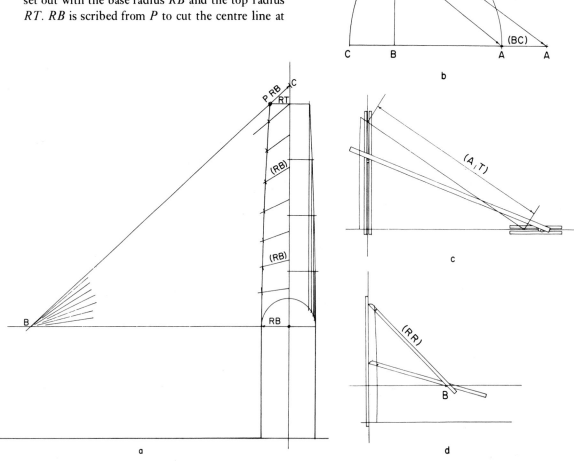

Fig. 50. Entasis (geometry): setting out

The right-hand side has the width of the column at the top dropped on to the bottom circumference. The arc caused by the diminish is divided into any number of equal parts. The height of the entasis is divided into the same number of equal parts. Those on the arc are projected vertically up to meet those marked horizontally across the shaft. Their intersection plots the entasis.

Figure 50 (b) shows the height of the entasis BT swung down to cut the base line at A. The top of the column T is then swung from A to cut the base line at C. The length BC is then added on to A to form A_1. The distance $A_1 T$ forms the trammel. The point T slides down the centre line while A_1 slides out along the base line.

Figure 50 (c) shows a trammel constructed to the setting out in Fig. 50 (b). The point tracing the entasis is on the rod $A_1 T$ with the width of the bottom radius outside T.

In Fig. 50 (d), a trammel is arranged as in the method illustrated on the left of Fig. 50 (a), tracing a full entasis. The end of the rod RR slides down the rule fixed to the centre line while the rod slides against the pin B. The point tracing the entasis is the radius of the widest part of the shaft marked in from the end of the rule.

EXOTHERM

See GLASS FIBRE.

EXPANDED METAL LATH

See BACKGROUND.

EXPANDED POLYSTYRENE

See BACKGROUND.

EXTRADOS

The outer, upper or back curve of an ARCH.

FAÇADE

The front elevation of a building.

FAIR-FACED

A term applied to smooth-finished concrete.

FALLS, LAYING TO

The slope to which a screed is laid that must carry water to drains, gullies, etc. See FLOOR LAYING AND ROOF SCREEDING.

FAN TEXTURE

See FINISH/RENDERING, EXTERNAL.

FAN VAULTING

See VAULTING.

FASCIA

A flat, broad, vertical member; usually with little projection. In classical terms a typical example is the wide member in the architrave of the entablature. In modern terms it could be the facing over a shop front.

FAT

1 The material brought to the surface when finishing in Class C or D plaster, or lime-setting stuff after scouring. This will not set and must be trowelled off. Hollows filled with this material will crack and flake on drying.
2 The residue left on the trowel when using water as a lubricant to obtain a smooth finish to other plasters. It must also be trowelled off.

FAT LIME

See LIME.

FAULTS

Faults can arise in all operations for many reasons but some are sufficiently common to be universally recognised. These are listed below.

Blisters and blubbing

The formation of convex swellings on finished plasterwork. It also occurs during working and may be caused by dusty backings or by air being trapped within the plaster, forming a bubble. At this stage the usual remedy is to prick the bubble or press the blister or blub back.

Bleeding

Water draining from a mix before it has hardened.
See also GELATINE.

Blowing

The blowing off of portions of a plaster coat due to the expansion of a material – ferrous metals rusting, CLAY balls expanding. See SAND.

Bond failure

The breakdown of adhesion between background and backing due to poorly prepared background, incorrect specification or poor application.

Case hardening

The hardening of the surface of a cement mix ahead of the rest of the mix underneath. It is caused by heat and, under the pressure of further working, results in cracking and sliding.

Cockling

See CASTING IN FIBROUS PLASTER.

Cracks

In fibrous plastering

Cracks on the face due to the expansion of insufficiently soaked timber reinforcement contained by the cast will take place immediately and, therefore, normally before fixing. Casts that are soundly fixed both to the background and to each other, with properly reinforced and keyed stopping, can only be made to crack by movement in the background.

In solid plastering

1 Pronounced cracks caused by: (a) Movement in the structure, causing the background to crack – due to settlement, differential coefficients of expansion (moisture and thermal) between different materials. Such cracks will often follow the structure, e.g.

stepped form along the joints in blockwork. Another cause is moisture movement in timber carrying lathwork. (b) Strong gauged cementwork on expanded metal lath. The mix shrinks and cracks on drying out. The stronger the mix the wider apart and bigger the cracks.

2 Fine hair cracks or crazing: (a) Usually caused by shrinkage of the undercoat after the finishing has set – most commonly occurs on cement-bound backing due to poor, clayey sand and/or applying the finishing before the backing has fully set or shrunk. (b) May also be due to shrinkage on drying of the finishing coat caused by the inclusion of too much lime. In lime setting, this type of cracking is avoided by the compaction resulting from violent scouring.

3 Fire cracking – with the use of pre-mixed, lightweight plasters this can take place in the undercoat as well as the finishing. It is caused by shrinkage due to the background or backing sucking too much water from the mix before it sets. In extreme cases the finishing coat will firm up and crack into areas 40 to 50 mm across. These curl up and drop from the backing.

Dry-out

The drying out of any freshly applied plaster before the final set can take place. Caused by suction, drying winds, sun, artificial heat or a combination of these and results in powdery or weak plaster.

Dusting

See FLOOR LAYING AND ROOF SCREEDING.

Efflorescence

A white, furry deposit on a finished plaster surface. It is caused by soluble salts brought by water from the plastering, background, or earth due to a faulty or non-existent damp-proof course. It is deposited when the water dries out. It will penetrate or push off paint, wallpaper, etc. As BS 1191 limits to a negligible amount the soluble salt content of plaster for application to backgrounds containing sufficient water to bring it out, troublesome amounts of these salts present in a plastering mix will usually have been introduced by the sand. Backgrounds of fired clay and clinker blocks can contain the salts. Once the structure has dried out completely, no more efflorescence can be brought to the surface unless dampness is introduced at a later date.

Failure to adhere to plasterboard

This is caused by: (a) Using hemihydrate plasters with excessive quantities of additional materials, e.g. the floating coat containing too much sand in two-coat work. (b) The use of anhydrous plasters next to plasterboard. (c) Using the wrong grade of pre-mixed lightweight plaster. (d) The inclusion of lime in the mix.

Failure between setting and backing due to lack of adhesion

1 With sand–lime or sand–lime–cement backing coats this is likely to be caused by: (a) Backing mix too weak. (b) Insufficient keying. (c) Backing still green when setting coat applied.

2 With gypsum backing coats this is most likely to be caused by a strong setting coat on a very weak backing coat. The weakness of the background may be due to one or both of the following: (a) Too much sand in the mix. (b) Soft sand.

Flaking

The peeling or scaling of the finishing coat of plaster. Due to lack of adhesion between coats through dust, lack of key or incorrect specification.

Grinning

The appearance on any plaster surface of the pattern of background joints and the like. It is usually due to variations in suction and can be overcome by applying a thicker backing coat.

Lime bloom

See LIME.

Pattern staining

The staining of a surface by dirt deposited on it according to the rate at which the structure conducts heat. The colder the surface the darker the stain. Therefore, when colder conditions exist behind the structure than in front, the thicker and more heavily insulated areas, such as wood joists and laths, will have their shapes silhouetted as lighter areas.

Popping

This can be seen on the face of finished plastering and is in the form of small conical hollows in the plaster surface. It is caused by the presence of reactive material such as unslaked lime. On becoming damp after the plaster has set, this slakes, expands and blows out the small hole.

Popping and pitting

See LIME.

Setting too quickly

When plaster sets too quickly it is usually caused by: (a) Heat speeding up the chemical reaction between the plaster and water. This may result when freshly made and still warm plaster from the manufacturers is used. (b) Dirt promoting the

formation of plaster crystals. The dirt may be introduced into the mix by dirty water; dirty, poorly washed sand; and dirty gauging board, tools, appliances, etc. (c) The introduction into the mix of partially set plaster from previous mixes. (d) Plaster stored in damp conditions and therefore partially hydrated. (e) The presence of clay with retarded hemihydrate plasters.

Shelling

See CASTING IN FIBROUS PLASTER.

Spalling

As Blowing (above).

Staining of solid plaster surfaces

This occurs when foreign matter that is present either in the background, e.g. rust, or in the backing, e.g. clay in the sand, penetrates through to the finished plaster surface.

Sweat out

The softening of plaster due to excess moisture. It may be caused by various conditions, one of which is the premature sealing of the plaster surface by an impervious paint while the plaster or its background is still wet.

Unsound

Term applied to a material or mix of which a part causes damage by expanding after it has set or hardened.

FEATHER EDGE

The tapering to nothing of the edge of a plastering coat.

FEATHER-EDGE RULE

See RULE.

FIBROUS PLASTERING

The work involved in the production, fixing and finishing of cast fibrous plasterwork.

FIBROUS PLASTERWORK

Fibrous plasterwork is plaster of Paris (Class A Gypsum casting plasters), reinforced with jute fibres. This current form appeared in the mid-nineteenth century with the introduction of jute woven into HESSIAN and enabled relatively large sections of plasterwork to be easily formed to an even thickness.

Cast fibrous plaster must be formed so that no reinforcement appears on the face of the plaster and it is capable of being fixed to a background. The face is formed by a layer of unreinforced plaster. Reinforcing RIBS across the back of a unit enable fixing by nailing, screwing or tying with wire and WAD and are also essential to the reinforcement of a unit as they impart rigidity. The ribs are comprised of wood lath or sometimes metal formers covered with hessian soaked in plaster. See also CASTING IN FIBROUS PLASTER.

Uses

As gypsum plaster is soluble in water and must be protected from the weather, fibrous plasterwork is mainly confined to indoor use. It can be thought of in two main forms: as a cladding material to line the interior of buildings; or as decorative areas supplementing other finishing materials.

As a cladding material fibrous plaster has many things to recommend it. It has all the advantages of gypsum plaster: FIRE RESISTANCE, ease of repair, and a final surface which is apparently joint-free. It will also receive conventional decorative surface coverings: papers and fabrics, paints – including those giving a grained or marbled effect, gilding, and thin-bed materials such as tiles and mirrors.

By casting from rubber moulds taken from natural materials (timber, stone, brick, etc.) it can, itself, have the surface texture of those materials.

The principal uses of fibrous plastering are:

1   Basic cladding on GROUNDS (q.v.) fixed to the structure of a building (and therefore following its shape). This would comprise PLAINFACE to walls and ceilings, BEAM casings, COLUMN casings, etc.

2   The cladding can be held away from the structure by fixing it to a grid. The space created behind the units can be used to house services such as air-conditioning ducting, electrical installations and plumbing. Provision can be made in the units for lighting or air conditioning grills and diffusers, down lighters, smoke detectors, sprinklers and surveillance and alarm systems.

3   The services in (2) can become features when designed to be so, e.g., concealed lighting by means of LIGHTING TROUGHS and reflector coves. These can be as naturally occurring features – CORNICES, beam case and bulkhead SOFFITS and column and stanchion casings to walls: or created features – the perimeter of FLOATS (lower areas of suspended ceilings), the perimeter of DOMES to ceilings and the perimeter of NICHES to walls.

4   To change the interior shape of a building by the use of, for instance, vaulting, domes and apses, so that an aesthetically designed interior may be housed in a structure designed on mechanical principles.

5   As embellishments, modern or traditional, used

in traditional positions: panelling to ceilings or walls, arcades, colonnades, CORNICES, FRIEZES, architraves, DADOS, SKIRTINGS, OVERDOORS, etc.

6 As imitations of structures formed in other materials: wood panelling or log cabins, bamboo and rush mat constructions (where fire regulations prohibit the use of the natural materials) and stone and brick buildings.

FIELDED WORK

Panelling – flush, raised or sunken. See PANEL, PLASTER (SOLID), PLAIN AND MOULDED.

FILLET

A small, flat member used in moulding design to separate other members. See MOULDING, SECTIONS.

FILLING IN

Synonymous with CASTING.

FINES

See SAND.

FINISH/RENDERING, EXTERNAL

All external finishes should be both durable and attractive and have some degree of resistance to moisture penetration – though this must not be taken to mean waterproof. A relatively absorbent rendering, which can 'breathe', is often the most satisfactory. A strong, dense mix is far more likely to crack; water will then penetrate the cracks, be unable to escape because of the denseness of the mix, and will travel inwards. This could cause dampness inside a building or continual dampness behind the rendering, eventually causing it to part company with the background or backing.

The materials used in external cement work are Portland cement (BS 12), hydrated lime (BS 890), sand – which should be both sharp and clean (BS 1199) – and coloured Portland cement or various pigments for colouring. Other aggregates used in obtaining textured finishes are shingle or crushed stone for roughcast and pea shingle or spar for pebble dashing. Also obtainable are several materials comprising PVA plus fine natural aggregates.

The backing is ruled to a true surface and keyed with a comb scratcher. Where the specification calls for three-coat work, the RENDERING is keyed with a comb scratcher to receive the FLOATING. The floating coat should be left at least 24 hours before the finishing coat is applied.

Cement plainface finish

This is probably the most common of all external renderings. It should be applied to a well-keyed backing or floating coat that is at least as strong as the finish. The usual gauge for this is either 2 : 1 : 8, 1 : 1 : 6, or 1 : 2 : 9, cement : lime : sand. Under certain conditions a 1 : 3 mix may be specified but this is usually considered too strong and dense.

The finish should be applied to the floating coat at an average thickness of 6 mm, pressed well into the floating with a laying trowel and laid to an even coat. Provided the floating coat is good and straight, screeds may not be necessary. A rule is then passed over the entire area, and depressions filled out and ruled off once more. When the coat has stiffened sufficiently, a 300 mm long wooden float is rubbed over the whole area until a uniform float-finish texture is obtained.

Where the finish has dried quickly, water should be sprinkled on the float face just before rubbing. Water should not be sprinkled directly on to the face as it will cause part of the Portland cement to be washed away, leaving a sandier face than is usually required. Flattening in with a laying trowel, although giving a pleasant appearance to 'green' work, will result in almost neat cement blotches appearing on the surface. If the required finish has not been obtained, a smaller float can be used. A flat sponge, sponge float or carpet float will not improve the flatness of the work, but provide a uniform texture, and these are infinitely preferable to a steel trowel. The material should always be applied with upward strokes, and a rule at least 1 m long used for ruling in. Bearing in mind that a wet cement–lime–sand mix should be used within two hours, it is, where practicable, a good idea to mix sufficient material to complete the job in one gauge. This ensures that the colour is uniform on drying. A scraped texture can be obtained by using a drag or old saw blade. Such a surface is less likely to craze and the scraping action will remove all surface LAITANCE. The resulting texture will depend upon the coarseness of the aggregate and the amount of scraping. In most cases scraping is left until the work has hardened – at least 24 hours.

Ashlar

This is cement plainface marked out in blocks by a jointer (see TOOLS) to represent plain masonry blocks. The chosen size of the blocks is marked out with a gauge rod while the plainface is still soft, a straight-edged rule is held in position and the jointer pressed into the surface and drawn along the edge of the rule, both horizontally and vertically.

Any selvedge is removed with a small wooden float. Overdoors, lintels, etc., are set out geometrically and occasionally a keystone is included. V-joints are obtained by fixing a V-shaped metal profile to either a wooden float or a piece of flat timber. It is nailed to the front edge and, with the bottom edge of the float on the rule and the V on the setting-out mark, the float is pushed along the rule. This must be done carefully: any attempt to obtain full V-depth at the first push will result in tearing. Bearing down should proceed gradually until the required depth is obtained. Any torn arrises can be touched up with a small feather-edge float.

English cottage finish

A well-keyed backing is essential. A thin coat of the finishing material is layed evenly over the backing. While it is still soft, small quantities of the same material are applied with either a laying or a gauging trowel in short, pinching strokes to produce a free-hand daub effect. No attempt should be made to obtain a regular pattern: casual effort will produce a better effect. Size, weight and spacing of daubs are variable but the best effects usually result from fairly close patterns and frequent overlaps.

Fan texture

About 10 mm of the finishing material should be laid on the well-keyed backing. Then the heel of a laying trowel or wood float is pressed into this wet coat, the tool being at an angle of about 45° to the horizontal. Keeping the heel in position, the tool is pressed into the surface, lifted, moved to the right and pressed in again. This is repeated several times to produce a fan effect. The tool is then lifted clear of the wall, and the process repeated at an adjacent spot. This finish is most effective in small areas. No attempt should be made to space or set out the pattern accurately. Overlaps are essential and at no part of the wall should a plain section be visible.

Imitation masonry

This finish can be confused with ashlar but imitation masonry usually consists of regular blocks with a rough-textured finish and deeper joints between the blocks. The area to be covered is floated to a good, flat surface and deeply keyed with a comb scratcher. If the joints are moulded, they are run with a running mould. If they are plain but deep and wide, suitable battens are prepared for fixing to the wall where the joints are required, then used as grounds for working to. They must be tapered to both sides – this will assist removal when the cement work has set. The run joints are formed before the finishing is applied. Long horizontal rules are fixed so that the cement moulding can be run in its correct position. On very large areas vertical joints are also run up, if possible at the same time, to form the mitres. On smaller areas it is usual to run lots of the moulding down before running the horizontal mouldings. These are cut to the correct size while still 'green'. On completion of each horizontal section, they are bedded up in position and the mitres cut into the horizontal moulding while it is still green.

When all the running, planting and mitring has been completed, the block faces are built out to the required thickness and the correct textures applied. The bedding of the vertical sections must be carried out in neat Portland cement. The sides of the run and planted section (usually straight to provide a ground for the finish) must be well keyed to ensure good adhesion of the finish. For running see RUNNING IN SITU, cement and sand.

Pebble dashing (dry dash)

This finish requires a different backing to that for most external finishes. As the name implies, the finish is produced by dashing or throwing on pebbles or spar to the final coat. It is essential that some control is exercised over the suction of the backing coat. This is achieved by including a waterproofer, either in the form of a liquid or a compound, or using a water-repellent Portland cement. The gauge must be 1 : 3, cement : sand, and the waterproofer added in accordance with the manufacturer's instructions.

The pebbles or spar should be clean, and free from dust. As preliminary preparations, as many clean pails as possible should be filled with pebbles and placed in a convenient spot; clean plastic sheeting, tarpaulins or clean sacks should be as near the area as possible. The mix for the final or butter coat is then prepared. This usually consists of a 1 : 1 : 5 cement : lime : sand mix; a colouring agent may be included. When the material is gauged it is placed on a spot board.

Pebble dashing is at least a two-man job and the two plasterers should start together to apply the material to the left-hand end of the wall. The thickness of the coat will vary according to the average size of the aggregate but must not exceed 10 mm and should be 1 to 2 mm less than the largest pebble or whatever. A rule may be lightly passed over the finish but this is not essential. While the material is still soft, one man drops back to do the dashing. He should first arrange the plastic sheet so that it will catch all of the aggregate droppings. Then, tucking a bucket of pebbles under one arm and using a dashing trowel in the other hand, he dashes pebbles on to the wet surface. By the movement of his wrist he should effect a wide spread of aggregate so that a uniform texture is obtained. As soon as he has caught up with his partner, he empties the pebbles caught in the plastic sheet back into buckets (they may need washing again), cleans up generally, and then starts on the fresh area that his partner has laid on.

After dashing is complete, a wooden float can be very lightly applied to the pebbles to press the aggregate back into the mix. This is not essential but dependent upon the type of finish required. Any jointing should be left irregular as it is then less noticeable when the work is complete. However, if

Table 2

FINISH Mix and thickness	REQUIRED BACKING Mix and thickness	DEGREE OF WATER RESISTANCE AFFORDED	BACKGROUNDS TO WHICH IT CAN BE APPLIED
Cement plainface (1) float finish Cement : lime : sand 1 : ½ : 4, 1 : 1 : 6, 1 : 2 : 9 6 mm (average) (and marked ashlar)	The same as the finish or stronger in cement content, never weaker. 9–15 mm keyed with comb scratcher	Weatherproof but not waterproof 1 : 2 : 9 unsuitable for severe exposure	Common brickwork Well-keyed and spatter-dashed concrete Moderately strong porous sand, lime and concrete blocks Lightweight aggregate blocks Woodwool covered by galvanised wire or EML
Cement plainface (2) float finish Cement : sand 1 : 3 6 mm	1 : 3 9–15 mm	Weatherproof but not waterproof unless containing a waterproofer	Strong, dense back-grounds only, well keyed and spatterdashed
Tyrolean	Applied to Cement plainface (1) not stronger than 1 : 1 : 6		
Pebble dash/Dry dash Mix depends on colour required – usually 1 : 1 : 5 5 mm for 5–6 mm aggregate	1 : 3 + waterproofing compound One coat 9–15 mm Two coats 9 mm each coat Both coats keyed with comb scratcher	Good. Backing must include waterproofing compound or liquid	Common and engineering brickwork Well-keyed and spatter-dashed concrete Moderately strong sand, lime and concrete blocks Not used on weak back-grounds, lightweight blocks, etc.
Roughcasting, harling, wet dash (1 : 1 : 6) 1 : 1 : 3/3 Cement : lime : sand/ aggregate 6–12 mm depending on aggregate size	1 : ½ : 4 1 : 1 : 6 9–15 mm Keyed with comb scratcher	Weatherproof but not waterproof	Common and engineering bricks Well-keyed and spatter-dashed concrete Moderately strong, porous sand, lime and concrete Lightweight aggregate building blocks
English cottage, fan and all heavy tex-tured work including masonry As for cement plain-face (1) and (2) ex-cept for 1 : 2 : 9 mix	As for Cement plain-face (1) and (2)	As for Cement plain-face (1) and (2)	As for Roughcasting
Natural coloured aggregate and resin finishes As manufacturer's instructions	1 : 3, 1 : 1 : 6, 1 : ½ : 4 9–15 mm float finish	Good, but not com-pletely waterproof	As for Cement plainface (1)

Note. Spatterdashing is recommended for all strong, dense materials, including engineering bricks.

possible, it is best to attempt all dashing in one go. For the safety of young children, it is advisable that the area of pebble dashing should finish about 1 m from ground level.

Instead of using a bucket as a pebble container, many plasterers construct dashing trays or boxes. These may vary in size from 450×300 mm to 600×400 mm and have three wooden sides 100 to 150 mm in height and a cross-handle about midway back. The sides may be tapered.

To form external angles it should be possible, provided that the floating coat has been well applied, to lay the butter coat either side of the angle and then dash in the normal way. However, the butter coat may be laid on one side to a hand-held rule and then dashed. The rule is removed and held gently on the side that has been dashed, and the operation repeated. Two points must be observed when using this method: the rule must not be pressed too hard against the dashed work; the two applications must marry at the angle.

Reticulated and vermiculated work

Reticulated means net- or mesh-like; vermiculated means worm-like. These types of finish are normally found only on keystones and quoin stones because, unless cast from reverse moulds as reconstructed stone, they take a considerable time to produce. A minimum of 6 mm thickness is built up and flattened with a wooden float. The shape and size of the pattern is then set out on the mortar and the pattern cut out. In some cases the reticulations and vermiculations are left proud, in others they are cut out. Originally, the intention behind vermiculated work was to give the appearance of a worm-made trail. More recently the finish has been used in reverse so that the vermiculations are made to represent the worm itself.

The tools used may be plasterer's small tools, knives, nails, filed strips of metal or metal rod. The areas between the patterns may be pricked or

pocked but plain bands of flat-float-finished mortar should be left around all edges.

Roughcast, harling or wet dash

In this finish, pebbles or crushed stone aggregate are mixed in with the cement, lime and sand, replacing half the normal sand content, e.g. a $1:1:6$ mix becomes $1:1:3:3$, cement : lime : sand : aggregate. A butter coat may be applied to the backing prior to dashing, to improve adhesion. Sufficient water should be added to the mix to make it wet enough to be thrown on to the wall. The container for the material (a hawk, a shallow box or a shallow, wide bucket) is held in one hand and the dashing trowel in the other. To prevent 'bunching' of the aggregate, as wide a spread as possible should be obtained: as the dashing trowel approaches the wall during the throwing operation the wrist should be used with an easy, flicking action. Bunching cannot be eliminated by spreading with the trowel after throwing – this will produce flat spots.

Work should be systematic, and as large an area as possible covered in one operation. The mix should be thrown with sufficient force to fill the keyings. Any joints should be left irregular. Noticeably smooth areas can be lightly brush-stippled with a similar mix at the end of the job but this should be unnecessary. External angles can be dealt with as for pebble dashing.

Tyrolean (machine finish)

This finish is normally applied to a backing that has been left as a cement plainface, wood float finish. A high-suction background is considered desirable for machine finishes and, therefore, the backing coat should not be stronger than $1:1:6$. A one-coat finish to blockwork, etc., is unsatisfactory as the Tyrolean process will not hide imperfections.

The mix may be of coloured Portland cement-sand, a ready-mixed coloured cement-sand, or a made-up pigment–cement–sand. The Tyrolean hand-operated machine is similar to a box with one end half-open. A spindle with a handle at one end passes through the box. Inside, spring steel strips are fixed to the spindle and these contact an adjustable trip bar when the handle is rotated, producing a flicking action.

The material should be mixed reasonably wet and applied to the background in three coats. Throughout the process, the operator stands about 0.5 m from the wall and for the first coat more or less at right angles to it. This first coat should be of a uniform honeycomb texture. When it is firm enough the second coat is applied, this time from an angle of about 45°. A uniform overall texture is essential: there must be no bald patches and no runs caused by excess material being applied to one patch. The final coat may be applied at an angle or directly on.

Coats should be staggered at the joints if these cannot be avoided but a better method of working is to apply the first two coats on one day and the final coat over the entire area on the second day. Windows and doors must be well masked out. The machine should be thoroughly washed out after use.

There are two refinements of this finish. The first is rubbed texture, obtained by going over the finished area on the following day with a block of carborundum stone or a block made of the same material as the finish. This produces a texture where all high points are rubbed flat but the indents are left untouched.

The second is scraped texture. In this case the area is scraped after 24 hours with a drag or old saw blade. The scraping should be just hard enough to remove the surface skin of mortar and expose the aggregate.

In both cases the degree of rubbing or scraping can be varied to suit the conditions.

Sgraffito

Scratched or incised work; light relief sculpture on walls. A strong hard backing of cement and sand is required and this must be well keyed to receive the following coat. The design is traced on this backing and then one or more coloured coats are applied in fairly small patches. While these coloured coats are 'green' the scratching operation can begin. The top coats are scratched away where required, leaving the colour necessary for the design. The backing must be perfectly hard and must set before subsequent coats are applied. When colours are used, the bottom coat should pick up before the application of further coats. Only small areas should be worked on and each patch should be finished before continuing to the next.

In modern sgraffito, areas of designs are masked out with tape, and the coloured material worked up to the tape. When setting is complete, the tape is removed and new tape placed over the completed work so that fresh colours may be applied to the plain areas. For this method the backing must be firm and hard but is usually unkeyed.

Natural coloured aggregate/resin finishes

There are several of these materials available today and basically all are similar. They consist of natural coloured aggregate and liquid synthetic resins. For the best results all require a cement plainface backing coat, and one make also requires a very light coat of its own undercoat to be applied before the finish. This finish can follow within a very few hours. All tools and appliances must be perfectly clean before these materials are used. They are supplied in airtight containers and, where possible, sufficient material for the entire job should be emptied out on to the board and knocked up at one time. Any surplus may be returned to the container which is then re-sealed. Both plastic and stainless steel trowels are produced for this work. The application is by steel laying trowel from left to right and as thin as possible. The material should be laid flat with no trowel marks left showing. It may need one trowelling with a damp, clean trowel as soon as it has been laid but no further working of the surface should take place. This may cause tearing. No water is needed when mixing the material, and no water should be applied to the finish coat. The backing should be clean and free from dust and may be brushed with a damp brush before application. Care must be taken to protect finished work from rain, and none should be carried out during frosty weather.

Coloured cement work

Most external finishes may be self-coloured. A ready-mixed coloured cement/aggregate material may be used, or pigments may be mixed with the Portland cement and, if possible, dry sand in the quantity required for the entire operation so that a uniform colour is obtained. A GAUGE BOX should be used for accurate proportioning.

A degree of suction control in the backing is essential; areas that dry out more quickly than others will usually dry out a lighter shade. Also, irregular suction will mean some areas requiring more attention with the wood float – again this will produce a variation in colour. Colours may be used in the finishing coat of cement plainface, ashlar, wet or dry dash and most other decorative finishes. Several manufacturers produce proprietary brands of colouring agents, and to produce the best results the maker's instructions must always be observed. Both powder and liquid agents are available; the liquid one should be added to the gauging water. The water content must be constant as it affects the final colour. For strong colours Portland cement and normal sand are used but for paler colours a white-cement-silver-sand mix is best. Lime bloom (see LIME) tends to give a faded appearance to paler colours.

Special materials

Alpine (white travertine) finish

This is a premixed finishing coat consisting of white Portland cement with graded aggregates. Surfaces must be primed with one part Cemprover to three parts water and the Alpine finish applied before the priming coat has dried. When properly spread with a trowel the material tends to 'chatter'. The travertine texture is obtained by dragging a wooden float across the face of the work.

Cemrend

A premixed one-coat rendering, applied by hand or spray machine. It is composed of white Portland cement, sands, light-fast pigments and additives to aid application, adhesion and weatherproofing. It can be used to produce all decorative finishes. Cemrend can only be purchased and applied by Snowcem recommended applicators.

Cullamix Tyrolean

This is a premixed finishing coat, consisting of white Portland cement, sands and pigments. It is applied by machine. See Tyrolean (machine finish) p. 96.

Dashing render

Is a premixed one-coat rendering to receive dashing and is based on Portland cement, lime, sand and light-fast pigments with additives to improve workability and adhesion.

One-coat system:	Apply 8–10 mm and allow to pick up then apply 6–8 mm to receive dashing
Two-coat system:	Apply 8–10 mm as a scratch coat. When cured for 3–5 days lightly dampen and apply 6–8 mm as a butter coat and dash.

Mineralite

Mineralite is a premixed finishing coat that gives the appearance of unpolished terrazzo. It is composed of white Portland cement, graded sands and aggregates that include granite and glass. Backing coats should not be weaker than 1:0.5:4. Mineralite may only be applied by Snowcem trained applicators.

One coat renders

Based on premixed Portland cement, lime and sand with additives to improve workability and adhesion, these are applied by hand. A lime-free version for tile backing is available. Three grades are produced: OCR1 (1:0.25:3) for dense, strong backgrounds and severe exposure conditions, 20 mm thick; OCR2 (1:0.5:4); and OCR3 (1:1:6), both for moderate/strong backgrounds and moderate exposure, 16 mm thick.

Scratch plaster

This is a polymer-based textured finish that is applied by hand or spray. It is an emulsion base with graded aggregates, fillers, pigments and biocides. Backgrounds must be primed with Snowcem GP primer. Spread the scratch plaster 1–2 mm thick with a trowel so that it chatters. Obtain the scraped texture by drawing a thin plastic trowel across the surface.

Note: All proprietary materials should be applied according to the manufacturer's instructions.

Miscellaneous points

It is almost impossible to cover every known external finish as there are so many. Various types of stipple finish, whorls produced by drags, brushes, rags and even fingers, brush curtain and drapery finishes, depeter (a form of dashing that has been pressed flat by a wooden float) and many others all go to make up a comprehensive list.

Bell castings are usually required over doors and windows, also where the external rendering finishes above another decorative feature. This is to direct water away from either the aperture or the feature.

The sand must be clean and sharp, the cement fresh and the water clean. No ordinary external finish should go below the damp-proof course as it will provide a way for rising damp. No external work should be carried out under conditions of frost and in hot weather one should 'follow the sun', i.e. work

in the shade after the sun has passed. Where possible, all work should be cured for at least one week after completion. See CURING, cement.

Scaffolding should always be of the independent variety. This will remove the necessity of making good put-log holes, etc.

FINISHING, INTERNAL (SETTING) TO PLAIN SURFACES

The finishing or finish coat is the final coat of plaster, applied to an internal surface. Its purpose is to provide a clean, smooth surface for decoration.

To obtain the best possible result with this coat, whatever material is being used, it is essential for plasterers to have the correct tools. These are a good laying trowel, a skimming float, a hawk, gauging trowel, small tool, water brush, angle trowels and crossgrain float.

The material to be used should be gauged correctly to a creamy consistency, free of lumps and in perfectly clean water. The backing to which it is to be applied should be straight, keyed by a devil float (except in the case of thin wall plastering), well brushed free of dust and if necessary lightly damped.

The whole operation of finishing is usually applied in three very thin coats, total thickness varying from 2 to 5 mm, depending upon the plaster being used and the backing or background.

Once the backing is ready and the material gauged, the first tight coat with the laying trowel may be applied. Whether to walls or ceilings, one should always work from left to right and in the case of ceilings, across the light. The material should be applied in long sweeps and every part of the backing covered, including all angles. A feather-edge RULE should then be passed over the whole area, and all irregularities removed or filled out with more plaster. This is essential in the case of all angles.

When this has been completed, the second coat is applied with the skimming float, again in long sweeps and leaving as little material on the surface as possible. The best way to do this is to apply a good quantity on the upward sweep and remove the excess with a returning sweep. The effect of this coat is to make the surface flat. The feather-edge rule may be lightly passed over the surface on completion, again to check for irregularities. When this is sufficiently firm to withstand light finger pressure, the third coat may be applied, again very thinly and in long forward sweeps. The reason for this is that by slightly overlapping the trowel sweeps, the toe of the trowel will remove the line left by the heel of the trowel.

When this is complete and the plaster is hardening, trowelling up may begin. A trowel is held in one hand and a water brush in the other. A little water is then applied to either the wall surface or trowel face; this acts as a lubricant, and the trowel is passed up and down the wall, or across the ceiling until a polished surface has been obtained. Two trowellings should be sufficient, one almost as soon as the last coat is complete, the second just as the plaster really hardens. To obtain good internal and external angles, the crossgrain float, lightly wetted, should be passed along all angles. All corners are then cleaned out and, if necessary, pointed in with the trowel end of a small tool; angle trowels, both internal and external, are then passed over the correct angles. A further trowelling may be necessary to remove any marks left by these tools, or the operations mentioned may take place between trowellings.

Finishing (setting) to rules

This would normally take place to window heads, window and door reveals and in certain cases to all external angles. The rules are fixed so that they project in front of the floating by the correct thickness for the finish. The plaster is applied as for finishing, ruled out with a reveal gauge or feather-edge rule and finished. When it has set hard, the rules can be removed. Sometimes the rule is then turned round and fixed to the set surface, in the same way as mentioned above but, more often, on removal of the rules, a slight selvedge will remain on the angle. This should provide a guide or screed for setting the adjacent surface to.

When this has set hard, the rule may be removed, or if the selvedge was used as a guide, all excess plaster cut back. A crossgrain float may be passed lightly over both faces of the angle if thought necessary, then the external angle trowel and finally the laying trowel, so as to remove any marks. With certain modern plasters this may not be possible and plasterers use many appliances to get the correct type of external angle-pieces of PVC, wash-leather, wet and dry glasspaper, even surforms.

Finishing (setting) to curved surfaces

On large curved surfaces the method of finishing is the same as for all plain surfaces except that, because of the shape, one must work with the curve; it is impossible to work from left to right. The curve prevents the centre of the trowel edge from contacting the backing, the plaster is applied in horizontal sweeps, starting at the top. Curved feather-edge rules are used to obtain a regular shape, and the system is as before: trowel, float, trowel. For smaller areas the finishing may be run. A pivot having been erected to carry out the floating, the same one may be used for the finishing.

As mentioned under FLOATING TO CURVED BACKGROUNDS, a muffle or batten was fixed to the rule to produce a backing, spun from the pivot. This batten or muffle is now removed and, when the rule is offered to the backing while still being attached to the pivot, it will be seen that a gap appears between the backing and the rule. This gap represents the thickness of the finish. The plaster is applied to the floating, horizontally from the top down and the rule is pivoted on the centre. This will form a near-perfect curve. After it has been built out so that it contacts the rule in all parts, the finishing may proceed either by trowelling immediately or by laying down and then proceeding to trowel to a smooth finish.

Finishing, one coat to plasterboard

For this operation a Class B, retarded hemihydrate board finish plaster is the one that should be used, Type b2. For all plasterboards other than plaster or gypsum lath, joint scrimming is essential. This is applied before the finishing coat and should be used at all joints including the wall/ceiling angles, although there are varying opinions on the usefulness of the latter.

A plasterboard ceiling should be finished before the walls are floated. A strip of plaster is applied across each joint, for the whole length of the joint and to a width of 100 mm, 50 mm either side of the joint. Hessian scrim, approximately 90 mm wide and which has been cut to the correct lengths, is then pressed into the plaster with the laying trowel so that it straddles each joint. At the wall/ceiling angle, 50 mm is applied to the wall, 40 mm to the ceiling. Scrim is also pressed into both sides of this, half on the wall, half on the ceiling. No overlapping of scrim should occur at any point.

When this has set, the area between the scrimmed joints are covered with fresh plaster so as to bring them out to the same thickness. Finally the whole ceiling is then given another coat of Class B board finish and ruled in straight with a feather-edge rule.

In most cases today, the craftsman will not use the skimming float on a one-coat finish to plasterboards. Having got the ceiling flat with a feather-edge rule, he will wait for it to 'pick up' and then lay it down with fresh plaster. As it begins to harden, it should be trowelled to a flat smooth surface with a minimum of water. Just a little, splashed on the trowel face to act as a lubricant should suffice. When the walls are floated, this coat will cover the ceiling angle scrim. (GYPSUM LATH, as for plasterboards with no scrim.)

Finishing in lime plaster (setting stuff and class A)

Originally this material was used neat on a raw lime coarse stuff backing. The suction from this backing was sufficient to give the craftsman all the hardening of the plaster he needed. Since then, however, gauged backings have become more common and this fact, coupled with the necessity for quicker-setting materials, resulted in the mixing of a quantity of Class A gypsum plaster with lime setting stuff. Quantities would vary, according to the backing suction and the finish. However, the application of the setting coat was as to plain surfaces.

Finishing, thin coat plaster

This plaster is a specially prepared retarded hemi-hydrate plaster and in general is made as a one-coat finish to most fair-faced backgrounds. It can be used directly on to concrete, both pre-cast and *in situ*, plasterboards and normal plaster backings.

As with all gypsum plasters, it should be gauged in a clean bucket using clean tap water. It should be applied in the normal way for setting, except for the skimming float coat which should be omitted. This method of setting is generally known as trowel and trowel, one coat being laid on, lightly ruled in and then the next coat laid down over it, again with the trowel.

As the plaster hardens, a very slightly watered trowel should be passed over the whole surface. Excessive water must not be used.

Finishing, using lightweight plaster

This finish should be applied to the correct type of undercoat before the latter has dried out. The first coat should be applied by trowel and pushed well into the keyed backing. A feather-edge rule should then be passed over the entire area, with great care being taken over all angles. Another tight coat is then laid over this as smoothly as possible, again with the trowel. Many plasterers dispense with the float coat in this case as the maximum thickness for lightweight finish should be 2 mm. As the plaster picks up, a light trowelling should be carried out. This should continue until a hard smooth surface has been obtained. Water may again be applied to the trowel to provide lubrication but this finish should not be overtrowelled. A smooth finish is generally more acceptable today rather than the hard polished surface required in previous times.

Finishing, plain surfaces, with the addition of scouring

This means that after the third application of the finish, the plasterer must sprinkle water over the whole area and scour hard with a crossgrain float, using a circular movement. As soon as one part of the wall has been scoured, and the scourer has moved along to another section, his partner would come along behind, trowelling off all surplus lime fat. When these two operations were completed, both men would go back to the beginning and trowel hard with water as a lubricant until a hard polished surface was obtained.

Finishing using either Class C or Class D plaster

These plasters are not currently commercially available but this entry may be of use in restoration work. With both of these plasters it is possible to obtain a very hard finish and, while certain arguments will be put forward against this, both can be finished in the same way. In both cases they should be applied to a hard mature BACKGROUND, using the three-coat trowel-float-trowel method. Also both

plasters will give better results both in appearance and durability if they are scoured, as above.

Some plasterers prefer to do their scouring after the second coat has been applied by means of the skimming float. When this has been accomplished, they gauge a little fresh plaster and lay down in the normal way.

Both plasters are at times capable of a quick initial set and both may be re-tempered, provided it is within one hour of gauging. Both plasters then have slow continuous sets and for best possible results should be applied early enough so as to finish before the end of the day. Keene's or Class D is an exceptionally hard plaster, ideal for squash courts, operating theatres and the like. Neither should ever be applied directly to plasterboards, or to weak soft backgrounds.

Finishing – faults

1 *Cracking* due to movement of the structure.
Remedy – cut out and make good with a similar material.
2 *Fire cracking*. This is normally the crazing of a finishing coat before the final set takes place and is due to extreme suction or poor application.
Remedy – damp the wall before setting and make certain that the initial coat of the finishing plaster is pressed well into the backing.
3 *Shelling*. Usually caused by poor adhesion between one coat of plaster and another. Dusty backings, poor keying, backing coats that are too weak will all contribute to this fault.
Remedy – make certain that the backing is the correct one for the finish before applying it. See that it is free from dust and properly keyed. A strong gypsum plaster applied to a relatively weak lime–sand backing will inevitably result in this fault occurring.
4 *Crazing*. Again caused by too much suction but when appearing on the face of finished work, especially Classes C and D plaster, it is more likely due to insufficient scouring or trowelling.

Remedy – make certain that all areas of work get the same amount of trowelling. Also endeavour to control the suction of the backing.

5   *Galls/cats' faces.* Caused by insufficient material being applied when laying down. They appear in the form of indentations on the finished surface.

Summary

As implied earlier, the best setting is usually carried out using the trowel-float-trowel method. However, much of today's finishes is applied by either trowel and trowel or float and trowel. This has its drawbacks as applying plaster with a float cannot always push the plaster into the keying and could result in *shelling*.

With well-floated lightweight plaster backings, the trowel and trowel method can produce excellent results.

FIRE HAZARDS See Appendix 1, p. 277.

FIRE RESISTANCE

For all details of plasters and fire resistance, BS 476 should be consulted.

FIRRING

1   The fixing of thin pieces of timber to the undersides of joists to make them level for lathing. A similar operation is carried out when fixing plasterboards to wooden bearers.
2   The wooden battens or metal supports fixed to a solid background in readiness to receive lathing in one form or another. The metal grid fixed to receive a suspended ceiling is sometimes referred to as firring.
3   A furry film of plaster coating a run moulding and formed by GATHERING ON.

FIRST AID See Appendix 1, p. 280.

FIRSTINGS See CASTING IN FIBROUS PLASTER.

FISHTAIL

A form of shoe used for spinning *in situ*, consisting of a metal strip with a V-cut in one end. The fishtail is attached to a gigstick, with the V-cut slots over a nail fixed at the centre point, and from this the running mould pivots. See ARCH, solid; FIXING, cement casts.

FISTING

The use of the fingers of one hand to feed with material as the mould is being run, thus assisting in the building up of members when running a moulding *in situ*. See RUNNING IN SITU.

FIXING

Fibrous plaster casts – to timber by nail or screw

Fixing to new work is normally carried out with galvanised nails through laths which have been specially placed flat in the cast and struck off. The nails are punched home below the surface of the cast and stopped in. Some specifications call for fixing with screws. Brass screws are infinitely preferable to the commonly used zinc-plated or dipped, as careless use of the screwdriver may strip the screw coating.

Old work, if sound, may withstand the vibration of nailing, but loose or suspect surrounding plaster will call for the use of screws – drilled and countersunk in the usual way. If nails or screws are not driven home to hold the cast rigidly to the fixings, the cast will subsequently move and the stopping be pushed off.

Fibrous plaster casts – to metal by self-tapping screw or wire and wad

Casts may be fixed to metal fixings arranged flat by screwing through flat wood laths or metal strips contained in the cast into the fixings, using self-tapping screws in holes that have been suitably drilled and countersunk.

The great majority of casts are fixed by wire and wad. This involves passing 18 g galvanised tie wire round the timber reinforcement placed on edge in the cast. The reinforcement is either wood lath in the cast or fixing battens wadded to the back. The wire is doubled and twisted. When wiring round the wood lath reinforcement, it is best to wire down through the cast, across the face in a countersunk groove, and back up the other side of the lath in order to fix round the full depth of the lath. The holes are drilled tight against the sides of the lath some 80 mm apart. Thus, each fixing employs a greater length of lath. The wire is passed round the metal fixing, twisted with a pair of top cutters and drawn tight by levering against the metal with the same action as that of withdrawing a nail. The ends of the wires are cut off, leaving about 15 mm of twist, and a juicy wad placed over the tie, following the wire down both sides of the metal on to the cast. The wads are rubbed on to the back of the cast to adhere and seek any undercut afforded by strike-offs or on-edge reinforcement. Wads hold the cast to the fixings, make the cast rigid with the fixings and prevent the wire from untwisting. (See Fig. 91.)

Miscellaneous

Cornices and other mouldings fixed to existing ceilings may need noggins placed between joists.

Supplementary fixings of metal or 25 × 50 mm timber may be wired across metal bearers on a metal grid where the join between two casts does not coincide with the bearers. The joint may then be regulated from the supplementary fixings by strutting down or wiring up. Where it is impracticable to get behind an existing hollow structure (such as plasterboard-, wood lath- or metal lath-faced surfaces) a dovetail hole is cut in the strike-off of the cast, filled with a wad that is passed through a corresponding hole in the fixing surface, and allowed to set. A variation is to fill each hole cut in the fixing surface with a wad.

When set, the cast is fixed by toshing a pair of screws or nails into each wad. Fixings may be made to strong, hollow backgrounds, such as clay pots, with expanding spring toggle bolts. If the whole bolt is passed through the cast, the large hole required will need to be bridged by countersinking a galvanised washer under the nut. Casts may be fixed to plastered brick and lightweight-block backgrounds with masonry nails or pins. Even galvanised plasterboard nails can be driven with ease into lightweight blocks and old stocks. Lightweight cornices that are small in section may be fixed by toshing short galvanised nails into the plastering to walls and ceilings. Such fixing may be reinforced by bedding the casts up. The strike-offs of the cast and the corresponding area on the fixing surface are best sealed with PVA. This will eliminate suction and improve adhesion.

Plaster of Paris is gauged to a creamy consistency, neat or with PVA in the gauging water. As it starts to pick up it is laid along the strike-offs in the cast, and the cast is then bedded in position and nailed immediately. The nails are punched home and stopped in with the same gauge. The plaster that has squeezed out at the wall and ceiling line is then cut off at just the right stage of its set to complete the stopping to the cast.

A cornice fixed along a wall running with the ceiling joints and having no fixing joist in the right place may be fixed by its cross-brackets or a special fixing batten positioned in the cast to coincide with a joist. The cast will require extra cross-brackets to reinforce it against the middle of the section being pulled back and to keep the ceiling line rigidly against the ceiling. The cast will be fully carried by these fixings but the ceiling line may have supplementary fixings of plug wads, toshed nails/screws, etc., in order to hold it to the ceiling.

Cement casts

Those features that do not lend themselves to being built into the structure or bedded as masonry can be fixed by dowelling with rustproof fishtails. The fishtail may be bedded in and left projecting from either cast or structure, then bedded into a dovetailed hole in the other with a strong cement mix; alternatively, both units may be formed with a dovetailed hole and both holes filled with a mix containing the fishtail.

Glass fibre

See separate entry for GLASS FIBRE.
    See also under specific items to be fixed.

FLAKING

See FAULTS.

FLANKING

Applying the floating coat between screeds.

FLASH SET

A quick and sudden set of gypsum plaster or cement on mixing with water. May be due to extremely fresh material or, more likely, to dirty water, old plaster or incorrect mixing. May also be caused by the adulteration of materials.

FLOATING

The backing coat that provides a true surface for the finishing coat. May be internal or external, on walls or ceilings, flat, curved or shaped. It must at all times adhere to the background or render coat and provide similar adhesion for the finish coat. It is carried out in numerous materials and, while there are small differences with each material, basically the application is the same. Thin coats are applied until the depth of the screed, angle rule or angle bead is reached, then ruled off to these grounds until perfectly flat and finally keyed for receiving the finish.

    Where areas have had screeds applied and the floating cannot be applied on the same day, the screeds must be trimmed straight. It will be possible to work on gypsum plaster screeds once they have stiffened (either through suction or set) but cement : lime : sand screeds are better left a while, otherwise they may be cut into by the floating rule.

FLOATING TO BEAMS

Horizontal crossmembers that project below the surface of the ceiling may be plastered with or without rules. In both cases the ceiling should be plastered first, at least up to the floating stage – this having been lined in to a datum line. If rules are being used they are usually wedged or nailed to the soffit first, so that the sides or cheeks can be plastered initially. To do this, fix one end first and check for square from the ceiling with a large wooden set square. Fix the other end, then line through with either a chalk line or long floating rule for straightness, and fix at regular intervals. The rule for the opposite cheek or side can be fixed parallel to this, using a gauge. When the floating coat is hard enough the rules are removed and work proceeds on the soffit. This time the rules are fixed to the cheeks, parallel with the ceiling (Fig. 51 (a)).

When working without rules, dots are squared in either side of the beam at each end (Fig. 51 (b)), checked for parallel with the gauge, then the intermediate dots are lined in from these; when firm, the floating is applied and ruled to the dots. The soffit is floated in a similar way, dots levelled in at each end, parallel to the ceiling, intermediate dots lined in or gauged down from the ceiling and then the floating ruled into the dots. In this case the external angles are worked in with a floating rule.

The setting or finishing coat is applied in both cases in the same way as the floating.

In large rooms or halls where several beams occur, these must be lined into a datum line so that all soffits are perfectly in line. The cheeks should also be checked for parallel with the aid of a pinch rod. In large halls where many beams occur, cradles can be used for floating the cheeks. Two long rules are fixed together by short cross-pieces, all to the same width overall. These are then wedged up under the beam soffit and squared in from the ceiling (Fig. 52). The term 'cradle' can also apply to the wooden members fixed at regular intervals along a reinforced steel joist to allow plasterboard or EML to be fixed around the beam. When wooden pieces are wedged in between the flanges of the steel they are termed 'noggin pieces'.

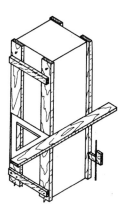

Fig. 52. Floating (solid): to a free-standing pier

Figure 51 (c) shows dots on a ceiling either side of a beam being levelled in with a levelling CRADLE.

FLOATING TO CEILINGS INTERNALLY

The operation of floating is fundamentally the same as that to walls. The same tools are required, and the same rules, though many plasterers prefer to use a box rule when floating ceilings. (See Fig. 57.)

See also DOTS, SQUARING IN; SCREEDS.

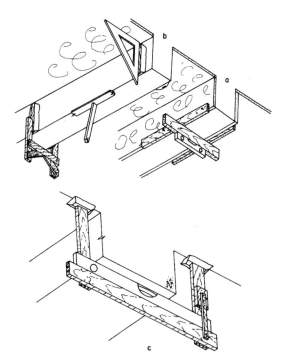

Fig. 51. Floating (solid): to beams

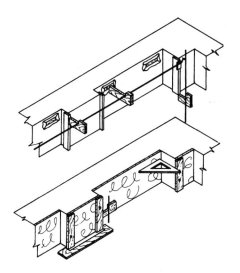

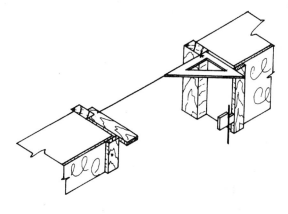

Fig. 54. Floating (solid): to square reveals

Fig. 53. Floating (solid): to walls containing piers

FLOATING TO CURVED BACKGROUNDS

Whether internal or external, to ceilings or walls, the basic principles for forming accurate curved surfaces are the same. (See Fig. 58.) One of two methods can be used, depending upon circumstances, or the size of the curve, and accurate setting out must precede the practical application of

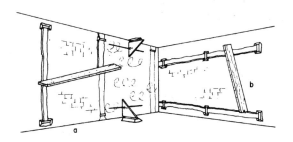

Fig. 55. Floating (solid): plumbing straight walls with dots and screeds

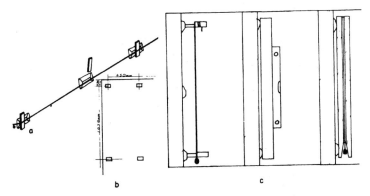

Fig. 56. Floating (solid): dots – lining in and plumbing

107

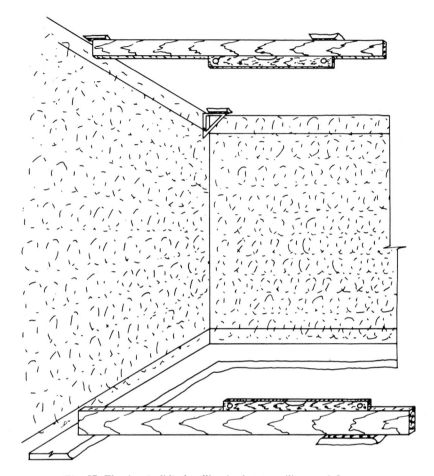

Fig. 57. Floating (solid): levelling in dots to ceilings and floors

material. The centres must be determined where applicable, and ribs cast or rules or templates cut to the line of the curve. DOTS are then lined in from the centre using a gauge rod, radiating from the centre of a line firmly attached to the centre point with the true radius measured on the line. At the end opposite to the centre, a short piece of batten is tied to the line. When this is stretched tight and the batten held at right angles to the line, dots can be ruled in position by this. Once the dots are firm enough, SCREEDS are laid between the dots and pressed in to form a true curve by the curved rule or template. This is the pressed-screed method.

The other method, used in general on smaller areas, consists of erecting a pivot, attaching an accurately measured rule to this and ruling in curved screeds by pivoting the rule on the central pivot – more or less in the same way as one would spin a circular moulding either on the bench or *in situ*. This system in fact is frequently referred to as the run-screed method (Fig. 64). Once the curved screeds are firm, floating can proceed as usual, the

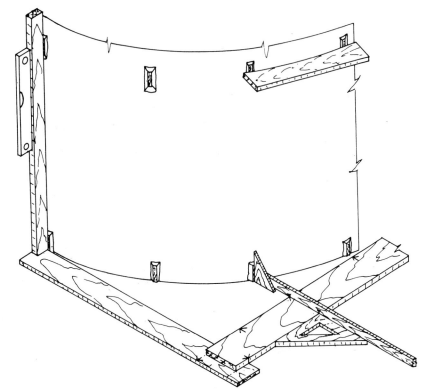

Fig. 58. Floating (solid): dots to a curved wall

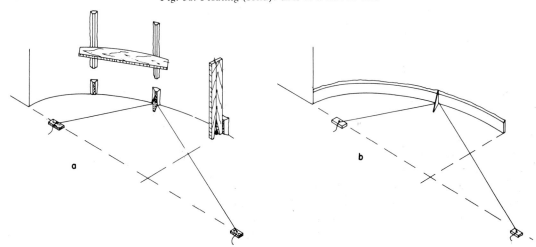

Fig. 59. Floating (solid): (a) positioning dots from focal points to a semi-elliptical wall, (b) wiping in a semi-elliptical screed from focal points

curved surface being formed by using a straight rule and ruling from the curved screeds. Where the work is circular in both directions, e.g. a dome or something similar, the curved rules or templates are used for ruling off the floating as well as the screeds.

Elliptical work can present additional difficulties, as centres cannot be used for these. However, once the major and minor axis positions have been determined, dots can be placed in these positions, and if necessary, mainly through size, intermediate dots gauged back from the axis (Fig. 58). Pressed screeds can be formed with the help of accurately made rules or templates from these dots.

Dots or screeds to elliptical work may also be ruled in from a trammel or (Fig. 59) from the focals. Should the work be internal and require finish, or if it is external and a cement finish is to be applied, allowance should be made for the extra thickness when working on the backing. This can be done either by adding what is virtually a muffle to the template or by having two sets of rules made, one for the backing, the other set for the finishing.

FLOATING TO FREE-STANDING PIERS

Especially if there are several independent piers in line, the cradle method is usually preferred for this operation. Several sets of rules are nailed together in pairs with cross-pieces, three per pair of rules. The cross-pieces should all be accurately cut to the pier side measurement, nailed to one rule so that the end of the cross-piece is flush with the working edge of the rule and then nailed to the other rule, again with the ends coming flush with the rule edge.

The two extreme end sets are fixed as for attached piers (FLOATING TO WALLS CONTAINING PIERS) and the other rule sets or cradles are lined in to these. When they are all fixed, two sides to each pier can be floated. Once these have set, the rules can be

taken down and the procedure repeated so that the other two sides may be floated.

This method should ensure that all piers are in line, provided care was taken at the setting-out stage and, with plumbing down, each pier should be square in itself. Pinch rods are used to check that the distance between piers is the same in all cases, also that the distance between all piers and nearby wall faces remains constant.

The same procedure should take place where the rules are replaced by metal angle beads. These should be lined in, margined and squared.

In all cases the floating material should be well worked in to the background, especially so at the external angles. Any loosely packed material applied here will either flake away on the removal of the rules, or crack and fall away at a later date. The material should, as ever, be applied in tight coats until the rule face depth has been reached. Then, using a small rigid floating rule, held against the thickness rules, and with the usual side-to-side action, the rule should traverse the entire height of the pier. Having completed opposite sides, remove the rules and repeat the operation on the adjacent sides. When these are complete and set hard, remove the rules, cut off any selvedge with a floating rule, and key with a devil float.

FLOATING TO REVEALS, DOOR, WINDOW AND CHIMNEY BREAST

The methods for floating all reveals are basically similar (Fig. 54). All reveals should have the same margin from top to bottom and be square to both faces of the floating or to one floated face and the constructional members, door and window frames.

In most cases it is essential that the projecting features should be floated first, namely the main walling or the chimney breast. Rules are then fixed plumb to this, so that they project in front of the reveal background by the floating thickness measurement. If the back wall has been floated, then a wooden square can be used, as in FLOATING TO WALLS CONTAINING PIERS, for forming the reveal. If the backing feature is a door or window frame then a reveal gauge should be made. This is done by holding a set square on the frame so that the adjacent side rests on the rule. A piece of wood is then laid on the square, flush to the front edge, and a mark made at the point where the frame ends. At this mark a nail is half driven into the edge of the wood. The square is removed and by resting the nail on the edge of the frame and traversing the reveal gauge from bottom to top of the frame a square reveal should be obtained. The frame margin will also be perfectly parallel, due to the margin on the gauge.

Faults

1 Cracking – due: (a) to movement of structure, (b) to poor application.
2 Crumbling.
3 Bad lines to finished work.
See also FAULTS.

FLOATING, WALL SCREEDS
TO RECEIVE HEAVY GLAZED TILES

The same method of floating should be used when one is preparing a wall screed to take heavy cement – sand, fixed glazed tiling. The backing for this type of work is normally a 1:3 mix, ruled in flat and keyed by wire or comb scratcher.

FLOATING TO WALLS CONTAINING PIERS

This can be a complex operation, with many things going on at once (Fig. 53). Provided each stage is carried out methodically the result should be perfection. The main problem with a wall containing several attached piers is to get all margins and measurements correct. There are several ways of carrying out this rather difficult job and the one we have used here is, we think, the best.

The first stage is to fix two rules, one on the left-hand reveal of the extreme left-hand pier, and the other on the right-hand reveal of the extreme right-hand pier. These rules must be the full height of each pier, must be plumb and should be fixed in such a way as to present sufficient thickness for floating the pier faces. A nail is then half driven in on the side face of each rule, approximately 150 mm from the ceiling, and the same operation carried out 150 mm up from the floor. Chalk lines are then tied firmly to both nails on the right-hand rule. These are then stretched tight across the faces of all the other piers until they reach the left-hand nails, and again pulled as tight as possible and tied to both nails. It should now be possible to fix side rules to all other piers, using the two lines as a guide. At the same time a shoulder gauge or gauge rod, with a saw cut to the correct size, should be used to measure distance between the two lines and dots on the back wall between all the piers.

The pier faces and the back wall can be floated at the same time, though – on the wall – one will need screeds ruled in to the dots. When all of this work has been completed the side rules should be removed and fixed to the pier faces. This time make a slotted shoulder gauge, the width of the slot between the shoulders being an accurate measurement of the pier width, or nail cross-pieces to the two rules removed from each pier, ensuring that the cross-pieces are accurately cut to the full pier width. They are fixed top, bottom and centrally to each pair of rules, the ends coming flush to the rule edges.

Once the rules have been fixed, the pier sides or reveals can be floated. Material is applied to each side, again in thin coats and then a wooden set square is held on the floated wall surface and pushed along with the right angle going into the floated corner and the adjacent side riding on the rule. The square is then moved upwards, bearing on one side on the wall, and on the other side the rule. Surplus material will be cut off and when the whole surface shows that it has been in contact with the square, the reveal will be square to the wall.

A good method of checking the squareness of the side with the face is to hold a rule horizontally at a convenient spot on the reveal; this rule must project beyond the external angle for at least 150 mm. Then, holding a set square with one side on the pier face and the right angle towards the rule, slide the square up to the rule. If every stage has been carried out correctly and checked for accuracy the adjacent side of the square should make contact with the entire length of the rule.

FLOATING TO WALLS EXTERNALLY

The main difference between internal and external floating is the type of plaster or mortar used. In Britain, gypsum plasters must never be used externally. All external work is carried out in Portland cement–lime–sand mixes other than waterproof renderings. With these, it is usual to omit the lime and replace it with a waterproofing compound or liquid.

In most cases, work is as for internal walls. The same tools will be required, though with most external finishes the floating should be keyed with a wire or comb scratcher (rather than a devil float) horizontally and in wavy lines.

However, on a strong dense background such as smooth concrete, engineering bricks and the like, it may be necessary to provide a key, and the method is SPATTERDASH. This not only provides a good key, but also regulates the background suction.

Once the dots and screeds are in position, the floating coat is built out in thin coats until screed level has been reached, then ruled off, till flat. Thin coats are less likely to crack or slip. Where the background is badly out of plumb, extra thickness may have to be applied. In this case the thin coats should be applied over a period of two or three days, each coat being scratched before the next is applied, again reducing the possibility of cracking or crazing.

When floating has stiffened it should be keyed by the comb scratcher in horizontal wavy lines. On completion, a short feather-edge may be passed lightly over the whole area to remove any projecting portions dragged up by the scratcher.

FLOATING TO WALLS INTERNALLY

Its main function is to provide a flat, plumb and true surface for the finishing. In three-coat plastering, it follows the rendering or pricking-up coat which must be well keyed to receive it. In two-coat plastering, it would be applied directly to the background which should be clean and free from dust, oil, etc.

See also DOTS; SCREED.

Tools required are a hawk and laying trowel, a lath hammer, a gauging trowel, water brush and devil float. The bays, or areas between screeds, angle rules and angle beads, are filled in by several thin coats applied by laying trowel. As the thickness of the floating attains the same depth as the screeds, anything from 8 to 12 mm (see BACKGROUND), it should be ready for ruling off. This is done by holding a floating rule across two screeds and at right angles to the screed surfaces, and bringing it up from bottom to top with side-to-side movements (Fig. 55 (a)). In the case of horizontal screeds the rule is worked up and down (Fig. 55 (b)). This action will cut off all surplus material, straighten the rest and show where additional plaster is required. A little extra is applied to areas that have remained untouched, and the ruling action repeated until the whole area is flat and true. The rule must be scraped clean after each ruling, and all surplus droppings picked up from the floor and returned to the spot board.

At this stage many craftsmen like to close in the floating with either a feather-edge rule or a darby. Neither of these two implements should be used in the first instance when obtaining a perfectly flat surface as they are usually slightly pliable. This can cause the floating to become concave or hollow. As the material begins to harden, either by the setting action of the plaster or because of background suction, it must be keyed ready to receive the finishing.

To do this the plasterer should lay the devil float flat on the wall surface and then traverse the entire area with circular sweeping movements with the float. All internal angles should be cleared out, by holding the laying trowel flat to the wall surface and diagonal to the internal angle, with the toe of the trowel just cutting the surface of the adjacent wall. The trowel is brought down from the top and the operation repeated on all walls at all internal angles. This should result in clean-cut angles that can be finished well. The gauging trowel is used to place floating material in place where it is impossible to use the laying-on trowel. The lath hammer is used for rule fixing and the blade end for chopping off projecting portions of bricklayer's mortar, etc.

FLOOR, CEMENT, GRANITE – MAKING GOOD

The following preparations are necessary to all types of flooring or screeding prior to carrying out repairs and making good:

1 All loose portions of the existing flooring must be removed with a minimum of vibration. Where cracks have appeared, the loose sections can be prised up with a cold chisel. Where no cracks are visible, but the floor has lifted, one sharp tap with a hammer at the highest point should be sufficient to produce the necessary opening.

2 The edges of the area to be made good should be trimmed to produce as regular a shape as possible. Sharp cold chisels and a bolster are essential for this.

3 Ensure that the edges of all existing work are sound and that there is no sign of bond failure.

4 Sweep clear all dust and saturate with clean water. Then just before laying, grout all surfaces to the areas, both horizontal and the vertical edges.

As far as possible the mix for repair work should be similar to the existing work, 1 : 3 for cement screeding and 2 : 5 for granolithic pavings. The new work should be tamped down to the level of the existing work and trowelled to produce perfect joints.

Surface repairs

These are usually carried out to cement screeding when the screed is to be covered by a thin-coat finish such as lino or thermoplastic tiles. There are several patent self-levelling flooring compounds available today for this, see LEVELLING COMPOUND, FLOORING. They can be worked to a feather-edge and should not exceed 6 mm in thickness. For small surface repairs where none of these materials are available a reasonable substitute is neat Portland cement mixed with liquid PVA adhesive. When well worked in to a clean damp surface, this will provide a good smooth surface but should only be used internally and when the screeding is going to be covered by another finish.

FLOOR LAYING AND ROOF SCREEDING

The laying of any cement–aggregate mix in the horizontal, to a surface that will probably be walked upon. All floor laying falls into one of two categories: the screed, topping or paving. While, in most instances, the procedure for carrying out both these operations will basically be similar, whatever type of floor, mix or finish, it is essential to understand the reason for the need for strict specification adherence. Far too many floors lift, crack, dust and in general cause unnecessary problems.

Floor and roof screeds are usually placed within one of the following categories of construction:

Monolithic. In this case the screed or topping is laid on a GREEN concrete base within a few hours, therefore bonding together and forming one unit. Minimum thickness, 20 mm.

Bonded construction. In some areas this is referred to as separate construction. It is used on an already hardened concrete base. This must be keyed and soaked, preferably overnight, before laying commences. Minimum thickness, 38 mm.

Unbonded construction. Here the topping is laid on either heavy waterproof building paper or heavy PVC sheeting. The thickness of the topping should never be less than 75 mm and the maximum bay size should be smaller than in monolithic or bonded constructions. The sheeting should overlap at least 100 mm at all edges.

Preparations before laying

Whether the floor is to be level or laid to falls, the first job is to strike a level DATUM LINE around the area or at least have levelled markers at suitable places. From this line, dots are levelled in with the aid of a wooden square. See DOTS, SQUARING IN (Fig. 57). Should the floor be laid to falls then a dot should be squared in to the lowest part of the floor and the remainder gauged back to the line so as to give a regular fall or incline. Great care should be taken, as errors could result in pools or puddles of water on the finished floor.

The second operation should be the setting out of the floor into conveniently sized bays for laying.

There are varying opinions on bay sizes but for monolithic and bonded floors, 15 m² is sufficiently large, we believe, and for unbonded floors 10 m². To assist in the splitting up of the floor into bays, and to give the plasterer a ground to which he can work for his levels, one of three operations now takes place. He will either lay narrow screeds, ruled in from his dots, he will set splayed wooden battens into beds of mortar or he will fix expansion joints – usually made of plastic – into the same position. The overall effect will be of a floor broken up into evenly sized bays or rectangles as on a chessboard, bays being laid alternately.

Immediately before laying – whether screeds, screeding or topping – the area about to be laid must be grouted in. Grout should consist of a neat cement–water slurry and this is brushed into the damp concrete surface. The floor must be laid on a wet grout. It may not be necessary to carry out this operation when forming a monolithic topping, but the concrete should at least be well brushed clear of LAITANCE.

Mixing is probably one of the most important exercises in the whole operation. For cement and sand screeding a semi-dry mix of 1 : 3 is normal unless the finished floor is to consist of a material that will be nailed to the screed, in which case it can be as weak as 1 : 5 or even 6. The quality of the sand in all cases is important: it must be clean, sharp, gritty and suitable for floor laying. For all granite work, a mix of 2 : 5 is usually considered to be the best, the granite chippings being graded up to 6 mm, with a small percentage of dust. Where the granite is either coarser than this or contains no dust, it is permissible to add no more than 20% sharp floor-laying sand. A line of the mix is then placed in the first bay, at the back or wall end; it is dropped in by shovel and left proud of the rules or screeds. A heavy, straight floating rule is then lifted above this and in regular movements pushed down on to the mortar until this is flush to the rules. This operation is called tamping.

When the first bay has been completed in this way, the rule is held on to the rules at the back of the bay and moved forwards in a side-to-side fashion. Then the area that has been laid is either trowelled once or floated with a wooden float once. The whole exercise is repeated until the bay is completed. Then it is usual to miss out alternate bays and to lay the next but one – this will eliminate cracking which is often caused by too large an area being laid at one time. At least 24 hours should elapse before the alternate bays are laid. In all cases the screed will require a further trowelling or floating. This should take place after the floor mix has stiffened and the surface can be brought to a good finish.

Screeds requiring a *float finish* are laid to receive mastic asphalt, asphalt tiles, emulsions such as bitumen, PVA and rubber latex, clay and concrete tiles, terrazzo, wood block and other relatively heavy finishes. Those requiring a *smooth trowel finish* are cork carpet and cork tiles, linoleum, hardboard and plywood, thin wood blocks, thermoplastic tiles, vinyl asbestos tiles and flexible PVC flooring.

Once the floor has hardened sufficiently, the wooden rules may be removed and the gaps filled in. In normal screeding these gaps may be filled by the same mix as the floor. In certain cases, finished floors will have their gaps filled, by bitumen or asphalt. All cement floors must be cured for at least seven days.

Granite toppings

The same operations preceding the laying are required for this type of floor – a level line, dots or rules, bays of a similar size and wet cement grout. The mix again should be semi-dry, as too much water in the mix can be one reason for dusting granite floors. There should be hard tamping down and ruling off and just sufficient trowelling to smooth out the rule lines. Alternate bays should again be the rule and if the joints are not to be filled with asphalt or the like, they should be removed and

the gaps filled the same day with the correct mix. A second trowelling is always necessary for granite, and this should take place later in the day when the granite is ready. For a non-slip surface, carborundum chippings are sprinkled on the granite just before the final trowel, usually 1 to 1.5 kg/m².

Curing

As with cement–sand screeding, a period of curing is necessary immediately following the laying of the floor. It should be kept moist for at least seven days – longer if possible. This will prevent moisture loss due to evaporation; damp hessian, polythene sheeting or wet paper are all aids in this direction.

Hardeners

Patent hardeners may be either added to the mix or applied to the finished floor surface, depending upon the manufacturers' instructions. A traditional hardener is silicate of soda diluted in about four times its volume of water. This is watered on to the floor surface and brushed in with a soft broom. The process may continue every 24 hours and for several days with the same mixture or one very slightly stronger. As soon as the surface has dried, all traces of the dried hardener must be swept off immediately.

Lightweight screeds

Portland cement and exfoliated vermiculite, usually 1:6. The vermiculite should be of the concrete aggregate grade and the advantages of this type of screeding are lightness, thermal insulation and some fire resistance. The mix should be thoroughly mixed together dry, by hand or mixer, water sprayed on gradually, then mixed for a short time; excessive mixing will cause compaction.

Lightweight screeding is usually thicker than other screeds and the laying should be done in the normal way. As soon as it has hardened it must be protected from wear by a coat of 1 : 4 cement and sand, approximately 19 mm thick.

Floor-laying faults

Dusting can be caused by many things – a poor mix, poor sand, a wet mix, overtrowelling and insufficient curing.

Lifting, caused by laying on a dusting or dry subfloor, a poor sandy grout or by laying bays that are too large.

Cracking, due to sub-floor movement, poor tamping and bad jointing, probably the joints being left too long before being filled in. Also shrinkage cracking can be the result of over-large bays.

See also LEVELLING COMPOUND, FLOORING.

FLOOR-LAYING AND ROOF-SCREEDING ADMIXTURES

These admixtures are usually in liquid form and, when added to cement screeding and granolithic paving mixes, accelerate the set and reduce the time of the final trowelling. They can also improve the workability of a mix.

FLOORING, TERRAZZO

A specialised floor finish consisting of marble chippings and Portland or coloured cement.

Under screed – see FLOOR LAYING AND ROOF SCREEDING, bonded or separate construction, float finish.

Preparations – mix proportions 1 : 3 cement/aggregate. The aggregate should be graded to contain three or four grades. The colour and cement must be mixed together thoroughly in a dry state. The aggregate is added, also dry and well mixed, water being finally added by a fine spray. The final mix must be plastic but not wet enough to spread or run. Floor thickness will be dependent upon the aggregate size but should not be less than 12.5 mm or greater than 25 mm. Jointing strips must be set down to form panels of 1 to 1.5 m². These may be brass, copper, ebonite or plastic and should be anchored to the wet screed.

Laying – the terrazzo must be laid while the screed is still green. If not possible, then a wet cement grout is essential.

The terrazzo mix is laid upon the screed, well tamped down, rolled and trowelled. Trowel once only, then leave for several hours before applying the final trowelling.

Stairs and steps should be treated in the same way. Tread thicknesses are recommended as 50 mm; risers and strings 30 mm.

Skirtings – as for floors. They should be laid to a cement wall screed and it is usual to complete the skirting before working on the floor.

Curing should commence from the time of the final trowelling until polishing begins. Wet sawdust, wet hessian or plastic are the usual agents.

Polishing may begin three to four days after laying. This is done by sanding and grinding with a coarse abrasive; the area is then washed, and filled with a neat cement mix, well trowelled in. The floor is then left for four to five days, still kept damp, then sanded with a fine abrasive. This procedure is continued until a perfect surface is obtained. Finally wash in a soft soap and hot water solution. Terrazzo floors are also laid in the form of pre-cast tiles, usually some 300 mm square. These are thick bedded in a cement and sand mix, grouted in and polished with a mechanical polisher lubricated with water.

FLUSH BEAD

See MOULDING, SECTIONS.

FLUTE

A concave groove. See also COLUMN.

FORMULAE

Circle

Circumference: $2\pi r$ or πd
Area: πr^2

Radius for an arc: $\dfrac{(\frac{1}{2}\text{ chord})^2 + \text{rise}^2}{2\text{ rise}}$

Circular dome

Area: $2\pi r^2$
Area of cap of circular dome: $2\pi rh$

$$\left(r = \frac{(\frac{1}{2}\text{ chord})^2 + \text{rise}^2}{2\text{ rise}} \right)$$

Ellipse

Circumference (close approximation):

$$\pi\sqrt{2(R^2 + r^2) - \frac{(R-r)^2}{2\cdot 2}}$$

Area: πRr

Parallelogram and trapezium

Area: half the sum of the two parallel sides × the perpendicular distance between them.

Triangle

Area: $\dfrac{\text{Perpendicular height} \times \text{base}}{2}$

'S' rule – area: $\sqrt{S(s-a)(s-b)(s-c)}$
$(s = \dfrac{a+b+c}{2}$, where a, b and c are the lengths of the three sides).

Pythagoras

The square on the hypotenuse of a right-angled triangle is equal to the sum of the squares on the other two sides.
$H^2 = a^2 + b^2$
$\therefore H = \sqrt[2]{a^2 + b^2}$
and $a = \sqrt[2]{H^2 - b^2}$

Cylinder

Curved surface area: $2\pi r \times \text{length}$

Barrel vaulting – surface area

Semi-circular: $\pi r \times \text{length}$

Semi-elliptical:

$$\frac{\pi\sqrt[2]{2(R^2 + r^2) - \dfrac{(R-r)^2}{2\cdot 2}} \times \text{length}}{2}$$

Regular polygons

Area: $\dfrac{\text{Number of sides}}{2} \times \left(\dfrac{\text{diagonal}}{2}\right)^2$

$\times \sin\left(\dfrac{360°}{\text{number of sides}}\right)$

Arc

Length: radius × the angle in radians

FRENCH CHALK

A kind of steatite used as a release agent for dry cement casting and to prevent the sticking of clay to surfaces during working.

FRESCO

The painting of newly applied lime plaster surfaces so that the paint combines with the plaster.

FRET

See ENRICHMENT.

FRIABLE

Crumbly and powdery. See STABILISING SOLUTION.

FRIEZE

The middle portion of the classical entablature. The area over the picture rail or architrave.

FROST

All new plastering and water-containing plastering materials must, before use, be protected against frost. New cement and sand should be protected for a minimum of 48 hours. Where work has been carried out internally in conditions of frost, this should be protected till dry. Any penetration by frost will cause ice to form within the plaster. When it thaws, the work will soften and peel away.

Work in pre-mixed plasters must take place where there is protection against freezing after application. As there is no sand content in these, they may be mixed indoors and, provided there is protection in the form of glazed windows and possibly some heating, work may proceed.

There are several admixes that may be added to either cement–sand or cement–lime–sand mixes that offer resistance to frost either by forming expansion chambers within the mortar, for ice crystals to form without disruption of the mortar, or by accelerating the set. Some of the latter also generate heat in the mix and this will offer resistance to frost attack, especially in the early stages.

GALL

See FINISHING, INTERNAL, TO PLAIN SURFACES.

GARGOYLE

A water spout, usually taking the form of a grotesque head and situated at the corners of buildings to throw rainwater clear of the buildings from roof gutterings, etc.

GATHERING ON

A furry deposit of plaster on the surface of a run moulding. It is caused by the profile lifting and not cutting the plaster off cleanly and to its full extent. See also RUNNING ON A BENCH, general method.

GAUGE

1 To mix.
2 A profile.
3 A rod marked or cut to size and used for measuring.

GAUGE BOX

A box consisting merely of four sides and used for proportioning mix materials by volume.

GEL COAT

See GLASS FIBRE.

GELATINE

The well-known substance that is produced by boiling butchery by-products.

Size

This can be produced by melting a little gelatine in warm water to such a strength that it will gel to a brawn-like texture on cooling. Setting is prevented by the addition of lime in the rough proportion of a gauging trowelful to a bucketful.

Melting

Gelatine is poured at blood heat in a molten state. It is melted in a water jacket, either in a special copper or one metal pail inside another. Oiling the containers before melting prevents the jelly from sticking. It is toughened by overheating and the melting temperature should not exceed 115°F. Fresh concentrated gelatine is best prepared by soaking in a small quantity of water until it softens, pouring off the surplus water, and melting; the water content is adjusted as necessary. For re-use, the jelly is cut into lumps, stirred during the melting process and removed from the heat as soon as possible. As with other hot melts, the cooler it is poured the better the chances of obtaining a good face on the mould. With open moulds it can be left in the bucket until a skin forms on top, whereas to ensure sufficient flow in enclosed case moulds it is best poured straight away.

Seasoning

Plaster and all porous models should be shellacked to a shine and oiled with boiled oil before pouring. Clay models are left to toughen and coated with thin coats of shellac, using a soft-haired brush, and similarly oiled. Gelatine is suitable for producing flood, clay-case, run-case and insertion moulds. For manufacture see MOULDS.

When cool and fully set, the mould is removed from the model, and any oil on the face of the mould washed off with turps (which is then mopped out with absorbent paper, etc.) Finally the mould is dusted with french chalk. The combined effects of water and heat from a plaster mix will destroy the gelatine; therefore the gelatine surface is 'pickled'. It may be sealed with a saturated alum solution which is either brushed over the face or poured into the mould and poured out again after some 15 minutes, then allowed to dry. The mould is then oiled before each cast. It may need re-coating with alum after some half-dozen casts. Alternatively the surface may be sealed with two coats of linseed oil varnish or lead paint.

Bleeding

Caused by a plaster cast being left too long in the mould. The heat generated melts any thin areas in the gelatine, causing it to weep through the sealer.

GEOMETRY

See ARCH; ELLIPSE; MOULDING, TO ENLARGE OR REDUCE; MOULDING, SECTIONS; PEDIMENT; POLYGON; RAKING; VAULTING, BARREL.

GESSO

1   A mixture of glue and linseed oil, stiffened with either plaster or whiting. Formerly used to form freehand-modelled enrichment.
2   Italian for plaster of Paris.

GIGSTICK

See SPINNING.

GIRTH

The measurement made at right angles across a moulding, following round the contours of its members.

GLASS FIBRE (WITH POLYESTER RESINS)

Seasoning for casting

Sealers

Porous surfaces (plaster, wood, etc.) from which a glass-fibre cast is to be taken are best sealed. This may be done with shellac or the resin manufacturer's proprietary sealer (cellulose acetate).

Release agents

1   Polyvinyl alcohol – water base, slow-drying, or solvent base, quick-drying. Non-porous surfaces (glass fibre, metal, glass, etc.) need only be coated with a film of polyvinyl alcohol. It may be sprayed, wiped on with a sponge, or brushed with a soft brush, and allowed to dry thoroughly.
2   Wax – on complex-shaped moulds, when maximum aid needs to be afforded by the release agents, the surface is coated with a silicon-free polish or with the resin manufacturer's product over the sealer if used, and polished before the polyvinyl alcohol is applied. The face of a cast can be produced polished by polishing the surface of the mould before casting.

Relatively small, single casts can be successfully released from a still-wet plaster mould, using only petroleum jelly as the release agent; very large casts (such as boat hulls) are released by the application of compressed air through lines connected at intervals to the back of the mould.

Casting materials

Surfacing tissue

This fine tissue is intended to reinforce gel coats and becomes invisible when wetted with resin but will render the resin translucent if many layers are used.

Cloth

Woven cloth – fine; woven rovings – coarse. These both provide glass strands running at right angles to each other and reinforcing the laminate in two directions only in the same plane.

Needleloom mat

This is basically a weave with strands looping up and down through the thickness of the mat, providing reinforcement in three dimensions through the thickness of the laminate.

Chopped-strand mat

The most commonly used glass-fibre reinforcement, consisting of flat, stiff, narrow ribbons like pine needles some 60 mm long and each made of numerous fine glass fibres. The glass strands are laid in all directions and held with an adhesive, providing two-dimensional reinforcement in the laminate. The adhesive is rapidly dissolved by the liquid resin so that the mat can be distorted by stippling with a brush to follow or fill the detail of the mould.

E glass

A chopped-strand mat is available with the same reflective index of light as a resin. Using this, the laminate appears as transparent as sheet glass.

Resins

Many types of polyester resin are available to suit various purposes, e.g. self-extinguishing (fire resistant), water resistant for boat hulls, highly chemical resistant, flexible, transparent.

The storage or shelf life of a resin can be from several months to a couple of years, depending on the type, the temperature at which it is stored (the warmer the shorter) and the amount of light (the more the shorter). It will, however, gel without the addition of any outside material. When the resin is required to harden, a catalyst is thoroughly mixed with it. The catalyst is an organic peroxide in liquid or paste form and should be used strictly in accordance with the manufacturer's instructions regarding quantities and safety. A catalysed resin will remain liquid for the best part of a working day. It is made to gel more quickly by the addition of an accelerator thoroughly mixed in according to the manufacturer's instructions concerning the working life required. The correct types of catalyst and accelerator should be used, according to the type of resin. Most of the more up-to-date resins are pre-accelerated by the manufacturers, according to the use intended. All mixing should be done with a stirring action, care being taken not to incorporate air in the form of bubbles.

Pot life

The chemical reaction caused by the accelerator gives off heat (the 'exotherm'). The more heat generated the quicker the gel and, therefore, the gelling of a thin laminate will take longer than that of a pocket of resin in a cast or in the pot. Resin spread too thinly and taking too long to gel will lose monomer vital to its hardening and so will not harden sufficiently. Draughts should be avoided. The temperature of the workshop will affect the pot life; insufficient curing will result from too cool an atmosphere. Pigments and fillers can also affect the gel time.

Additives

CEMENT-FACED CASTS

Dried cement or cement and sand can be mixed with a resin to the consistency of a paste and applied as a gel coat. Such casts will weather to be visually indistinguishable from cement and sand casts.

COLD-POUR METAL

Up to 80% of metal powder can be added to the resin to produce an imitation of metal (copper, brass, aluminium, etc.). The casts may have a polished finish or be oxidised with an oxidising agent.

FILLERS

Powders are available for mixing with resins to thicken them to any consistency up to that of putty.

PIGMENTS

These are in the form of pastes for mixing with resins to produce an opaque or transparent result. The manufacturer's instructions are best followed as certain substances will retard and weaken the cure.

Casting by hand – plain areas

The mould surface is brush-coated with a layer of resin. This is termed the gel coat. It forms a face to the cast that is free of glass fibre and protects the fibre itself from wear and moisture attack. The thixotropic gel coat will cover a mould evenly. It is inadvisable to have areas of resin not reinforced with glass; the essence of the structure is to have the glass fibres that provide the tensile strength coated with only enough resin to make them rigid and bind them together, rather than a volume of resin containing fibres. The necessary layer of unreinforced resin facing a cast should therefore be kept to a minimum – around 0.3 mm thick.

When this has hardened but still feels slightly tacky the laying up may commence. Woven rovings and needleloom mats need to be cut to size and shape beforehand but chopped-strand mat is best torn as the frayed edge feathers away to nothing, leaving no ridge on the back of the cast. The lay-up resin is mixed and sufficient brushed over the gel coat to receive each piece of glass mat. The mat is laid on the resin and pushed into it to make it ooze up through it. This is done with a hard stippling action by a brush. It may then be rolled with a spacer roller. The glass must be thoroughly wetted by the resin with no air pockets or bubbles in or under the mat. (Resin to impregnate the glass should not be applied on top of the mat as it cannot be thoroughly worked through to bond with the layer underneath or easily be made to impregnate the layer being applied.) The cast can be built up to the required thickness by impregnating successive layers of mat in this way. Details of resin : glass ratios are available from the manufacturers.

Reinforcing ribs may take the form of timber lath, battens or metal sections laid over the back of the cast and incorporated into it by covering with laps of glass mat pre-soaked in lay-up resin. Paper rope can also be used in the same way to shape the laps that, when set, themselves act as the reinforcing ribs. Ribs should not be applied too soon to a plain surface as the shrinkage of the resin within them can pull the cast round them, causing a bump on the face of the cast. Casts in the round hold their shape fairly well but, in general, flat casts tend to twist, due mainly to the shrinkage of the resin on curing. Ribs can counteract this to a great extent.

Glass fibres cannot be turned in and struck off. The most efficient way of finishing the edges is to produce a cast slightly full and trim it to size after removing it from the mould. The mould may be made slightly larger with a line incised to provide a trimming line on the cast, or the cast can be made to overlap the edges of the mould and trimmed off back to the strike-off line. 'Green' casts are easily cut with a trimming knife or even by tin shears, but those in a more advanced stage of cure will have to be cut with a cutting/grinding wheel or fine-tooth saw.

Casting by hand – moulded and enriched sections

This requires all the techniques of casting plain areas and, in addition, overcoming various further difficulties: (i) Coating the face with an even thickness of gel coat which tends to collect in arrises. (ii) Filling the detail in fine enrichment with neat resin is not so detrimental in an ornament as in a structurally functional item but thick pockets can create excessive exotherm which may cause them to split and crack and melt moulds of hot-melt compounds. (iii) The glass fibres must enter the detail and follow the contours, not cobwebbing over them to leave air pockets behind the gel coat and cause shelling (see FAULTS).

The best glass for filling detail is chopped-strand mat. The liquid resin quickly dissolves the binder, fixing the strands, and the mat can be made to stretch locally by stippling with a brush so that moderate contours can be filled. Deep, pronounced enrichment and arrises, however, can best be lined by pulling tufts from the mat and forcing it into the resin-coated detail with the stippling brush. Reinforcement can consist of timber or metal, positioned as for CASTING IN FIBROUS PLASTER and covered by resin-soaked glass-fibre laps.

Fixing

Cast-to-cast

This is done by lining up the casts and wadding them together across the back along the length of the joint. First, resin is brushed on to each cast some 50 mm each side of the joint and 100 mm wide strips of glass mat soaked in resin are placed over the joint and stippled with a brush. If necessary, splints of timber or metal may be placed across the joint and covered with resin-soaked glass-fibre wads in the same way.

Cast-to-background

As it is a disadvantage to pierce and make good the face, fixing from behind is preferred. It is made in much the same way as for the cast's counterpart in plaster or cement. Screw or wire fixings to the background are covered by resin-soaked wads of chopped-strand mat. Fishtail metal dowels contained by the cast may be embedded in glass-fibre wads in fixing holes in the background.

Stopping

The technique is to fill the joint with a slight excess of thick resin that does not run. This is rubbed down flush when it has hardened sufficiently. Special resins are available for stopping, or casting resins may be thickened to a paste with filler powders. Rubbing down may be done with a surform, file, power grinder/polisher, abrasive papers, etc. See also Appendix 1, p.279.

GLASS REINFORCED GYPSUM (GRG)

Glass reinforced GYPSUM is a material comprising gypsum casting plasters reinforced with GLASS FIBRES. It is a development of FIBROUS PLASTERING.

It is known as glass reinforced *gypsum*, rather than glass reinforced *plaster*, in order that its initials, more commonly used than the full term, are not confused with GRP, the accepted designation of glass reinforced plastics.

The plasters used are special hard casting plasters. Glass fibres impart to these plasters far greater tensile strength per thickness of lamina than do the jute fibres of fibrous plaster. Compared with fibrous plaster, where the lamina is an in-fill between CROSS BRACKETS and supports, GRG can be thought of more as a rigid self-supporting unit with provisions for fixings to hold the unit in position. The cross-sectional shape of the unit will dictate the amount of bracing and support it will require. Like the egg, some shapes are naturally strong: DOMES, VAULTING and those containing ribbing as part of their design. At the other end of the spectrum, the weakest shape is the flat, horizontal PLAINFACE.

With some forms of fibrous plastering the time spent forming the RIBS or cross-brackets to reinforce a cast can be as long as that taken in forming the basic lamina. When fewer brackets are required time is saved in casting, cutting and preparing materials and fewer materials are used. As with fibrous plastering, the brackets used are designed in conjunction with the grid to afford both reinforcement and fixing points.

Moulds

GRG is extremely hard and therefore far more difficult to clean up than fibrous plastering. With the latter, a scrape over with a joint rule or piece of BUSK and a rub-down with a canvas pad will remove grease and mould marks. With GRG, however, an impression can only be made with a joint rule or busk if they are sharpened for each cast and the operator puts in a lot of hard work. These considerations would therefore dictate that moulds for casting GRG should have faces of a high quality. Nevertheless, as with most plastering situations, no firm rule always applies. If only one or two casts of a particular section are required, spending extra time on cleaning up a cast taken from an improvised mould may produce a cost saving. On the other hand, if the number of casts is sufficient, it may be cheaper overall to produce a glass fibre mould so that no cleaning up of casts is required.

Production

With certain forms of glass reinforcement, providing the face is not to be scraped in the course of cleaning-up or STOPPING, the glass can be allowed to penetrate through to the face, as it will not be visible when the cast has dried out. When it is essential that the glass reinforcement is kept back from the face, the firstings and seconds method of casting can be employed. In instances where, for example, the cast is to receive light scraping, such as that received accidentally during stopping in, the reinforcement can be kept back from the face by first applying a surfacing tissue.

The lamina may be produced by two methods:
1 Cast from moulds with the reinforcement in sheet form, in the same way as fibrous plastering.
2 Cast from moulds by spraying. The plaster mix is sprayed from a gun with an overhead glass chopper which cuts continuous glass rovings; these fall into the plaster stream as used in the technique, with traditional overhead chopper guns, for the production of spray-applied glass fibre laminates.

GLYPH

Vertical channel in a TRIGLYPH.

GOING

See STAIRCASE WORK IN GRANITE AND CEMENT.

GRANITE CHIPPINGS

See AGGREGATE.

GRANITE FLOOR

See FLOOR LAYING AND ROOF SCREEDING.

GREASE, PLASTERER'S

Used as a release agent for plaster on metal, wood, etc., and all surfaces sealed with shellac. Also as a lubricant for the bearing surfaces of running moulds.

In emergencies, cooking lard or petroleum jelly will serve as excellent release agents. The traditional plasterer's grease, however, is based upon tallow with a thinner oil added to produce a consistency that may be brushed or spread with a rag over a mould, as a thin film, without obliterating fine detail. A fine balance must be maintained as a grease of too thin a consistency will be rubbed off by the brushing in of the firstings. Cooking oil is the preferred oil for this purpose.

The proportion of tallow to oil will vary according to the season of the year, as ambient temperature will affect the consistency of the grease. For example, when casting on a hot day with a special hard casting plaster gauged to a thick consistency, tallow alone may be used and requires no more than the softening effect of holding in the hand, whereas casting in the depths of winter calls for a grease with a high proportion of oil.

The tallow is melted slowly in a metal container over a low flame, care being taken to observe all the safety precautions necessary when heating fat. When all the tallow is melted it is *removed from the flame* and allowed to cool a little if it has approached boiling point (a state at which it will give off choking fumes). The oil is poured in and the whole stirred thoroughly.

Greasing up

The quickest and most efficient method of applying grease is with a rag, except on fine, deep detail where only the hairs of a brush will reach. The grease rag is a pad of canvas impregnated with grease. Hard bits of plaster, etc., picked up by the rag may be removed by opening it out and shaking it before each greasing operation. Although the rag can apply the correct amount of grease to plain areas, it may leave a deposit in arrises and other detail. This build-up is removed with a grease brush, the plasterer using this as a scoop and forcing the hairs forward. The grease so removed is wiped from the brush with a rag after each stroke. No grease marks – brush marks, streaks, etc. – should be visible. The surface should have been given a uniform shine by the grease and when all possible has been wiped off with a greasy rag or brush the correct amount will be present.

In fibrous plastering, plasterers use grease as a barrier cream and release agent during all operations that involve working plaster directly with the hands. However, the grease is not applied to hands until after the initial gauging operation, while the plaster is soaking.

GREEN

The state of a material at the latter stage of its set or hardening.

GREEN SUCTION

See SUCTION, green.

GRID, FOR FIBROUS PLASTERWORK

Two types of gridwork can be designed for fibrous plasterwork: one to which the casts are fixed by wire and WAD and the other to which the casts are fixed by self-tapping screws. Each system has its own advantages and disadvantages.

Fixing by wire and wad

When fixing by wire and wad the reinforcement in the cast is positioned on edge to offer the strongest fixing to the wire. Reinforcement on edge also means that the cast will be more rigid and will span greater distances before it requires support, in general a support bar every 450–500 mm. In addition, extra reinforcement, such as metal channel members, can be incorporated in the casts so that they can be more self-supporting and require fewer grid members. The casts will be suspended a little way under the grid on wires and their heights can be infinitely adjusted by raising and lowering the wires. The casts can be held rigid by inserting wooden wedges between the STRIKE-OFFS in the cast and the grid as the wires are tightened or slackened off. The cast can also be regulated in the horizontal plane by sliding the wired cast along the grid before tightening the wedges. This adjustment in the position of the casts means that the grid need only be constructed within a 10–20 mm tolerance. Therefore with this method, all positioning, alignment and levelling is accomplished by wire adjustment and wedging from the grid. See also FIXING.

Fixing by self-tapping screws

When fixing by screw the reinforcement in the cast will need to be positioned flat to present a strip suitable for receiving screws. This means that the cast will be less rigid than if the reinforcement were on edge and will need supporting more frequently, in general every 350 mm. The cast will be screwed tight against the support bars. This means that the accuracy of the fibrous plasterwork, in positioning, alignment and levelling, will rely on the gridwork being accurately set out and erected. This method would naturally be favoured when access to the back of the grid is not possible, such as if the grid is too close to the background or if the void behind is filled with ductwork and other services.

Advantages and disadvantages of fixing methods

With the screw method the grid takes more time to construct but the casts can be fixed rapidly. Conversely, the grid for wire-and-wad takes less time to erect but fixing the casts takes far longer. Where

total precision is called for, such as installing many casts end-to-end in a straight line, the fine adjustment possible in wire-and-wad is essential.

Erecting the grid

The full-size drawings of the sections of each cast will contain its supporting gridwork, shown in position against the back of the cast. The position of each member in the grid will be dimensioned from the face of the cast.

Sequence of operations

1 The positions of the hangers need to be marked on the background. This can be achieved by first establishing the building's grid and datum lines. Then, using the layout drawings, the lines of the fibrous plaster casts are set out. These lines can be snapped on the background. The position of the lines of hangers can now be marked parallel to the line of the casts and snapped. The position of the fixings (maximum 1.200 m centres) can be marked along them. The type of fixing can vary from patent attachments to existing metal, to forms of expanding bolt or anchor in a hole drilled in concrete backgrounds.

2 The fixings can now be positioned.

3 Straphangers are cut from continuous bars. Each has a hole made in one end, which is bent at right angles. They are fixed to the fittings through the hole and bolted in position.

4 The underside of the primary bar is dimensioned from the face of the cast. Its height can, therefore, be measured from a datum line and marked on a hanger. This mark can be levelled round the other hangers by means of a laser or with a water level and snapchalk lines.

5 Using a mechanical bender, the hangers are bent up to receive the primary bar at this mark. This procedure ensures that these bends are all in line in a horizontal plane, regardless of irregularities in the background.

6 The primary bars are laid in the bends and wired in position. These bars may be used alone to support the plasterwork, but where a secondary grid is used they should be no more than 1.200 m apart. Alternatively, the hangers can be cut off to length and bars fixed to them with rivets or nuts and bolts.

7 The channels forming the secondary grid can now be wired or bolted in position under the primary bars.

Cross-bracing is fitted where necessary within the grid and/or to the background to ensure that the grid is rigid.

GRID, FOR SOLID PLASTERWORK

See also Background, PLASTERBOARD. A grid will, in general, consist of three main components (Fig. 59A).

1 Metal strap hangers, 25 mm wide by 3.5 mm thick.

2 Main bearer channel, 38×12.7 mm by up to 1.5 mm thick.

3 Firring or secondary channel, 25×25 mm, by from 1 to 1.5 mm thick.

The hangers are fixed in line at 1200 mm centres, with the rows staggered. If to concrete, they are either built in during erection or bolted in position afterwards. The end of each hanger is bent up to form a cradle to receive the main channel and is wired back to itself above the channel with 18 g or 1.219 galvanised tying wire. To this frame the firring or secondary channel is wired, again with 18 g or 1.219 tying wire at 300 to 325 mm centres. Metal lathing is then wired to this as in metal lathing, steel fixing (see BACKGROUND, lathings).

Other types of grid fixings are metal rod hooked or stapled to wood joists, wooden hangers bolted or screwed to joists and metal rod bolted to steel beams. In the place of metal channel, angle iron may be used or alternatively wooden main bearers and firring pieces. The spacings will remain constant and the size of the component will depend upon the size and weight of the finished ceiling.

However, in general today the majority of grids are formed by the use of the channel or angle-iron bearers and metal strap hangers.

See also PLAINFACE.

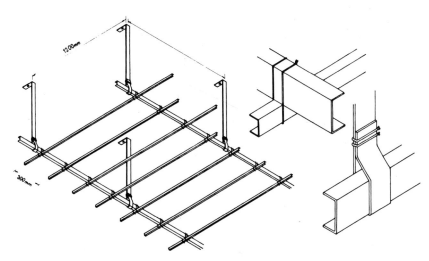

Fig. 59A. Ceiling, suspended (solid): metal grid

GRID LINES

See PRODUCTION MANAGEMENT.

GRINNING

See FAULTS.

GROIN

The mitre line produced by intersecting vaults. It may be left as a plain arris or covered by a moulding.

GROUND

In fibrous plastering, a special member carried by mouldings and off which panels are ruled in and to which they are finished. Used chiefly in the formation of MODELS and MOULDS. In solid plastering, wooden battens fixed to a background. The thickness should be the same as that of the plaster specified. The ground is then used as a screed for working to when ruling in the floating coat.

GROUND, FIXING

Timber batten, usually fixed to a solid background, to which fibrous plaster CASTS are fixed.

GROUND, MOULDING

The surface adjacent to the edge of a MODEL that produces the STRIKE-OFFS on a mould cast from it.

GROUT

A mixture of cement and water brushed into a surface to improve adhesion immediately before the application of a cement and sand mix. See FLOOR LAYING AND ROOF SCREEDING; WATERPROOFING, internal.

GUILLOCHE

See ENRICHMENT.

GYPSUM

Calcium sulphate dihydrate, $CaSO_4.2H_2O$. The mineral from which gypsum plasters are manufactured. See also PLASTER.

HACKING

Used as a means of keying concrete in readiness for plastering; may be carried out with a hammer or a power tool. See also SPARROW PECKING.

HAIR

The hair of animals such as goats, oxen or cows is used to reinforce plastering mixes. It was used extensively in the lime plastering era but many substitutes are now available, i.e. sisal, tow and man-made fibres.

HANDRAIL

Usually cast for external use in GLASS FIBRE or CEMENT mixes. For production see MOULD, REVERSE, plaster moulds – run slip-piece and the material in which it is to be cast.

HANGING MOULD

A running mould horsed in such a manner that it hangs on the running rule. Used to form moulded BEAM sides and SOFFIT members. (See Figs 10 and 62.)

HARDENER

For cement see FLOOR LAYING AND ROOF SCREEDING.
For plaster see DEXTRENE.
For resin see GLASS FIBRE.

HARLING

See FINISH/RENDERING, EXTERNAL.

HEALTH AND SAFETY see Appendix 1.

HEATING PANELS, CEILING AND WALLS, PLASTERING TO

Heating panels are grids of small-bore hot-water pipes invisibly incorporated within the structure. The surface heat should not be allowed to exceed 105°F (40°C). For the specification purposes, the panels are classed in two categories:
Embedded, in which case the panels are within the surface or soffit.
Suspended, part of a suspended metal lath ceiling

Heating panel, embedded

The surface under the pipes will usually be either clay filling and panel heating tiles, or concrete. In the latter case, the key must be equal to that provided by the clay tiles. These are single cast units of the clay tiles, listed under BACKGROUND.

Lime plastering specification two-coat work, floating and finishing, total thickness 19 mm. The mix for the floating coat should be 1:3 soaked hydrated lime : clean sharp pit sand; 4.5 kg/m² of well-beaten hair, manilla or sisal must be added and distributed evenly throughout the mix. This is then mixed with Class A plaster to a ratio of 1:2 plaster : coarse stuff. Size may be added to retard the set.

The floating coat must be applied in two coats of equal thickness and all work completed in one day, no re-tempering is permitted so that it can be finished without delay. The first floating coat must be cross-scratched and the second keyed normally for setting.

The finish is in lime setting stuff, 2:3 soaked hydrated lime : clean washed sand. This is mixed with Class A plaster to a ratio 1:2 plaster : lime setting. Surface scrim is not necessary where the panels are cast *in situ*, provided the key for the undercoat is adequate. (See below, heating panel, suspended ceiling.)

Pre-mixed lightweight plaster specification – for this a render coat of Portland cement and haired coarse stuff 1:4 is necessary, applied to a maximum thickness 9 mm over the entire ceiling and cross-scratched to form an undercut key. It is then left for 7 to 10 days. The floating coat must be premixed lightweight bonding plaster applied to an average thickness of 6 to 7 mm and lightly keyed for setting. This must be applied after a lapse of 24 hours and the total thickness of both coats should be no greater than 9 mm.

Heating panel, suspended ceiling

These ceilings require the application of a *pugging* mix. This means embedding the pipes in gauged haired coarse stuff all round except for the top surface, which will be covered by cork slabs.

Before pugging from below, loops of 16 g galvanised tying wire should be placed over the pipes at 150 mm intervals. The mix of Portland cement : coarse stuff 1 : 4 is vigorously applied to the area between the pipes in three coats. When complete, expanded metal lathing 20 g 6 mm mesh, is wired tightly to the pipes, using the loops of tying wire. All joints in the lathing must overlap by 150 mm. The render coat is gauged to the same proportions as the pugging mix and must be applied over the wire to a thickness of 6 mm while the pugging is still wet. When firm it must be cross-scratched and left till dry.

Lime plastering specification – the mix for the floating and finish will be as for heating panels, embedded (above). In the finishing coat, surface scrim is necessary, and dry 3 mm mesh hessian canvas scrim is applied to the face of the setting coat when it has been levelled and laid down. The scrim must extend for 300 mm beyond the edges of the heating panel and be trowelled into the surface. No plaster should be applied over the scrim; this will flake off. The scrim should be just visible in good light at a distance of 600 mm.

Pre-mixed lightweight plaster – pugging and rendering as for lime plastering. The area surrounding the heating panel should be rendered and floated in lightweight metal lathing plaster. The heating panel area must be floated as in embedded panels, using lightweight bonding plaster. 100 mm wide scrim with 6 mm mesh should be trowelled into this coat at all edges where a change of material takes place. Finish in lightweight finish plaster and the combined thickness of the two coats should not exceed 9 mm.

Application of heat

Heat must not be applied till the plaster is dry. During the first week the water temperature should not rise above 80°F (27°C), the second week 100°F (38°C) and for the third week 130°F (54°C).

HELIX

One of the spirals or small volutes found under the abacus of a Corinthian, Ionic or composite capital.

HESSIAN

A jute weave of 3 mm and 6 mm square mesh, used to reinforce fibrous plaster casts. The 100 mm width for reinforcing the plaster across the joints in plasterboards is termed SCRIM.

HINGE MOULD

See COLUMN, spiral; NICHE, SOLID; RUNNING, DIMINISHED.

HINGED GIGSTICK

See SPINNING, circle spun on a drum with a changing curve.

HORIZONTAL MOULD

See NICHE, SOLID.

HORNS

Protuberances on the back of a CAST, formed by covering the knuckled joint between reinforcements (metal or timber) with soaked canvas laps.

HORSE

See MOULD, RUNNING.

HUNGRY

A sand or a mix that has poor workability and continually requires knocking up with additional water.

HY-RIB

See BACKGROUND.

IMPOST MOULDING

A moulding round the top of a pier to support an ARCH, etc.

INSERTION MOULD

See under MOULD, REVERSE.

INSULATION, THERMAL

In plastering, thermal insulation can be improved in the following ways:

1 The use of plasterboards that have one side faced with polished aluminium foil to reflect heat within the cavity. Therefore when fixed to wall battening they will increase the insulation considerably.
2 All cavities provide a degree of insulation, so even normal plasterboards and other lathings fixed as in (1) will reduce heat loss.
3 Thermal board can be used, this consists of gypsum plasterboard bonded to a backing of insulating foam.
4 Wood-wool slabs and expanded polystyrene are both good insulating materials, see BACKGROUND.
5 The use of pre-mixed lightweight plasters also reduces heat loss. These plasters have approximately three times the thermal insulation value of sanded mixes.
6 Lightweight roof screeding will assist both in thermal and sound insulation.

INTRADOS

The inner curve of an ARCH

JESMONITE COMPOUNDS

See Appendix 4, p.294.

J-FOAM

See Appendix 4, p.295.

JOGGLE

A depression and corresponding protrusion, hemispherical or angular, for accurate location of pieces of a reverse mould.

JOINT, EXPANSION

A joint used (a) between backgrounds having different rates of expansion or situated where movement within the structure may take place; (b) between different materials used for surface finish; (c) between bays in large floors where expansion may take place within the topping.

In walls, where the background is involved, the joint may consist of a polyethylene strip inserted into the background at the junction of the different materials. It is then used as a ground for plastering to. Alternatively, a joint may be left and, when plastering has been completed, filled with twin-pack polysulphide. Where different wall plasters are being applied, a casing bead or plaster stop may be fixed at the joint prior to plastering. This type of joint is also used between plaster and wood or plaster and metal.

In floors, expansion joints may consist of bitumen-impregnated composition boarding, ebonite or plastic strips, in which case the material is generally used as a ground to the floor as well, or of hot- or cold-pour bitumen, which is applied after the floor has been laid, gaps having been left.

K LATH

See BACKGROUND.

KEY

The surface indentations which, when filled with plaster, serve to bond the two surfaces together. Natural keying is provided by the joints in brick and block work. Artificial keying is provided by wood and metal lath, by scratching and grooving block and brick during their manufacture and by scratching undercoats before or after they have fully hardened.

Plaster key

The portion of a plaster mix that penetrates or passes through the key afforded by the surface to which it is applied.

KEY PATTERN

See ENRICHMENT, fret.

KEYSTONE

The block, usually tapered, at the centre of an ARCH. For model see BRACKETS.

KILLING

Destroying the set of gypsum plaster by over-re-tempering or knocking back.

KNOCKING BACK

See RE-TEMPERING.

KNOCKING UP

Mixing materials.

KNUCKLE JOINT

See VAULTING, BARREL, casting.

LAITANCE

This is usually found on the surface of concrete sub-floors and consists of a thin layer of water and fine particles brought to the surface by excessive working or a too high water content. It must be removed by a wire brush before plastering.

LAPS

Small square, rectangular or triangular pieces of canvas used when casting, either to cover the cast as a part of the main layer of CANVAS – particularly useful with awkward shapes – or to cover the wood or metal reinforcement in a cast. In the latter case they are usually soaked in plaster before being applied.

LATH TANK

A narrow, shallow trough, some 300 mm wide and 200 mm deep, containing water in which the long laths of a cast are soaked prior to casting. Its length is determined by the length of the laths to be soaked. Conventional tanks are made specially to order by welding sheet iron and are better for being galvanised or painted to combat corrosion. A GLASS-FIBRE tank may be cast from a single run mould of which the sides and ends should be slightly splayed to allow the cast to be drawn off the mould. A makeshift tank can be produced from 150×25 mm planks of wood for the sides and bottom, with squares of the same for the ends. A single sheet of polythene is folded to line the interior, and the whole filled with water.

LATH, RIVEN

See BACKGROUND.

LATHWORK

See BACKGROUND.

LAY-UP

See GLASS FIBRE.

LEVELLING COMPOUND, FLOORING

Thin, smooth-finish material that can be applied to uneven and rough floor screeds prior to laying lino tiles, thermoplastic tiles and other thin floor finishes that require to be laid on a near-perfect surface. Two types are available: the *thin-bed* which are basically Portland cement plus powdered PVA and are applied to clean, damp floor surfaces, trowelled in tightly and trowelled again to finish; and the

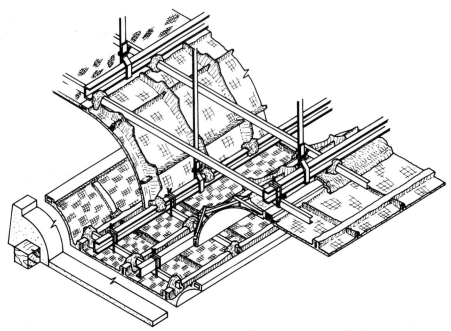

Fig. 60. Lighting troughs (fibrous): a trough and flush reflector coves fixed to a metal grid

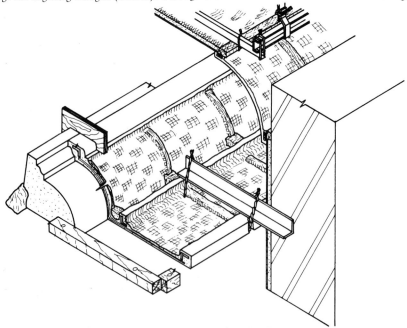

Fig. 61. Lighting trough (fibrous): a trough and reflector cove fixed to a wall

134

thicker-bed which are composed of pre-mixed dry sand and cement plus powdered epoxy resin, which so decreases the viscosity that they become self-levelling.

LIFTING-OUT STICK

See PLAINFACE, casting.

LIGGER

See SPOT BOARD.

LIGHTING TROUGH

A length of plasterwork, shaped in section to house and conceal the source of electric lighting. It may be double-sided when forming a soffit to such features as BEAMS and double-sided ARCHES (Fig. 60), or one-sided, serving as a CORNICE-like moulding with a SOFFIT cantilevered from the wall or when forming the edge to such feature as suspended CEILINGS and DOMES (Fig. 61). The trough is used to best effect in conjunction with a reflector cove.

Fibrous

Moulds required for the one-sided trough are of the type used for cornice work, whereas those for the double-sided trough are of the type used for BEAM CASES.

Casting

The trough is cast as a BEAM CASE but may require extra reinforcement, depending on the support afforded by the fixings. Where the trough is required to withstand cantilevering under its own reinforcement, metal cross-brackets are often used. Metal will also bend to the section, giving support to the riser while keeping the trough thin and hollow so as best to accommodate the electrical fittings. The inside edge of the top of the riser will cast a shadow on the ceiling and should be run up to a smooth edge by striking off with a gauge run off the top of the riser on the mould.

Fixing

Lighting troughs are usually fixed to metal. For procedure see PLAINFACE, fixing to metal.

Solid

Solid plastering lighting troughs are usually run off *in situ* using, in the absence of a nib bearing, a twin-slippered running mould (Fig. 62). They are formed on a metal lathing background, the lathing being fixed to brackets that may be either rod, channel or angle iron let into the wall at 300 mm centres. The running rule is fixed so that the uppermost of the two slippers will bear on to it. At the nib end of the metal profile a 25 mm top member should be included. Then, when the trough has been run off, wet plaster may be applied to the back of the moulding and ruled off to the top member with a wooden gauge. The moulding must be cored out before running with a plaster–sand–hair mix and finished in either lime putty and plaster or Keene's cement. See RUNNING IN SITU.

LIGHTWEIGHT AGGREGATE

See AGGREGATE, perlite and vermiculite.

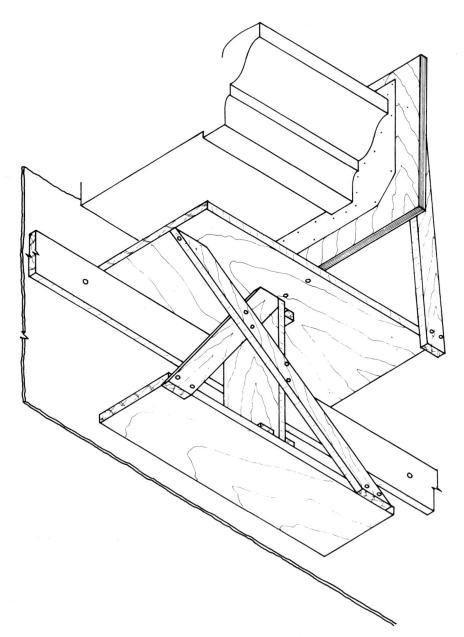

Fig. 62. Lighting trough (solid): a trough being run with a twin-slipper hanging mould

LIGHTWEIGHT AGGREGATE-PLASTER

See PLASTER, GYPSUM.

LIME

The term embraces many types and forms of lime. The one used in plastering mixes is hydrated, non-hydraulic lime in the form of lime putty or, mixed with sand, in the form of coarse stuff or setting stuff. It is mainly produced from calcium carbonate as limestone or chalk. The calcium carbonate, $CaCO_3$, is broken into lumps and heated in kilns to drive off the carbon dioxide, CO_2. This leaves lump quicklime, CaO. When cool, this is sprayed with water which reacts violently with high-calcium lime, giving off heat, expanding the lime and breaking it down into a fine powder – hydrated lime, $Ca(HO)_2$. However, the feebly slaking hydraulic limes will need to be ground to a fine powder. Another, though minor, method of production is as a by-product of the manufacture of acetylene gas.

Non-hydraulic

The near-pure calcium carbonate limestone and chalk will produce a non-hydraulic hydrated lime. This does not set or harden by reacting with water but, on drying out and combining with carbon dioxide in the air, reverts to calcium carbonate:

$$Ca(HO)_2 \nearrow^{H_2O} + CO_2 \rightarrow CaCO_3$$

Hydraulic

Hydraulic lime produced from calcium carbonate that contains clay, introducing silica, alumina and iron as impurities, is similar to a natural cement, having a set that will take place on reacting with water.

Semi-hydraulic

Limes having a lesser set than that of the fully hydraulic limes.

Blue lias

Lime produced from limestone that contains clay; it is therefore hydraulic.

Greystone limes, grey limes

Limes made from chalk that contains a small amount of clay; they are therefore semi-hydraulic.

Magnesian limes

These are made from dolomite limestone. This is a mineral composed of calcium carbonate and magnesium carbonate, $CaCO_3$ and $MgCO_3$, and, although most are non-hydraulic, a few are semi-hydraulic.

White limes, mountain lime, chalk lime, high-calcium lime

Limes made from limestone or chalk of practically pure calcium carbonate – non-hydraulic.

Terms

Dolomitic limes – see magnesian limes, above.
Fat lime – a non-hydraulic lime of good workability and high-volume yield.
Free lime – lime brought to the surface of a plaster mix, probably forming a lime bloom. Free lime can also be present in a cement that is produced by burning calcium carbonate.
Lean lime – a hydraulic lime of poor workability and low-volume yield.
Lime bloom – a film of hardened lime on the surface of plasterwork, produced by hydrated lime being

brought out and deposited by water in the course of drying out.

Popping and pitting – miniature craters in plasterwork, caused by unslaked lime particles in the mix eventually slaking when in contact with water, expanding and popping off pieces of plaster. To prevent this fault (apart from using a known sound lime) the lime should remain soaking as lime putty for as long as possible and then be passed through a fine sieve.

Putty lime – hydrated lime made into a putty by mixing with water.

Slake – to add water to quick lime to produce dry hydrated lime or, with an excess of water, lime putty.

Soundness – a sound lime does not expand after it has been used. The most common cause of expansion occurring in unsound lime after use is the presence of unslaked particles as described above for popping and pitting.

Volume yield – the increase in volume of the lime when quick lime is slaked to produce hydrated lime.

See also Appendix 1, p.279.

LIME PLASTERING

Coarse stuff/lime mortar/lime haired mortar

A traditional plastering material which, although not used extensively today, can still be found on some sites. Its function is that of a backing and in the period of lime plastering it was the only material used for this purpose. Generally, lime plastering consisted of three-coat work, render and float (in coarse stuff) and set (in setting stuff). Hair was always used in coarse stuff. It acted as a reinforcement to the mortar which, pushed through the KEY on lathwork and allowed to dry, would be almost impossible to remove. Usually it was added to the coarse stuff at the same time as the water, during gauging, and was in the ratio of 15 kg/m³. The mortar was allowed to mature for many weeks before use, in the form of alternate layers of slaked lime and sand or a mixture of these substances. When required, it was pulled down by a labourer, turned over and then ringed out to receive water. At this time the hair was added.

In present-day practice, coarse stuff may be delivered to the site as a ready-mixed lime/sand mortar or gauged on site, using sand and hydrated lime in a ratio of 1 : 3. If 'hair' is required, sisal, tow or a man-made fibre is used.

Setting stuff

A mixture of three parts lime putty and two parts fine washed sand.

Raw lime plastering

Solid plastering carried out in raw lime coarse stuff and setting stuff. It is usually applied in three coats – render, float and set – each coat being allowed to dry out thoroughly before the next is applied.

Render coat

9 mm thick and cross-keyed with a lath scratcher.

Floating coat

Ruled in flat between SCREEDS and keyed with a devil float. Average thickness, 9 mm.

Setting coat

Applied in three coats – trowel, float and trowel-scoured and polished. Average thickness, 3 mm. With no set to take place, the suction afforded by the floating is more than adequate to stiffen the setting. The coat must be thoroughly compacted by scouring with a crossgrain float lubricated with water, otherwise shrinkage of the lime will cause crazing. Sufficient fat is worked up by the scouring to provide a smooth surface when trowelled with a laying-on trowel. For proportions see above, Coarse stuff and Setting stuff.

Gauged lime plastering

As for *raw lime plastering*, with the addition of either Portland CEMENT or a Class C gypsum plaster to the backing, giving a 1 : 1 : 6 mix, and Class A, B or C plaster added to the finish in a ratio of 1 : 1 to 1 : 4 plaster : lime. This eliminates the waiting period normally necessary in raw lime plastering. It also makes the render coat unnecessary, except when applied to wood or metal lathing.

LINE, STRUCK OR SNAPPED

A dusted chalk line held lightly to previously applied marks at either end of a plain surface. The line is drawn away from the surface at the centre and at right angles. It is then released and allowed to snap back against the surface, leaving a coloured line.

LINING IN

Ensuring that a number of features (beams, piers, etc.) are all in line.

LINTEL

The horizontal member that spans an opening in a building. It may be of wood, metal, concrete or stone and may be the finished article or may have to be covered with plaster.

LIPS

Ridges initially formed in a core in case moulding. They will ultimately be on the back of the layer of pliable moulding compound, locking it into a corresponding groove in the plaster case.

LOOSE-PIECE

See under MOULD, REVERSE.

LUNETTE

1   A semi-circular feature in a ceiling and containing ornament – modelling, painting, etc.

Fibrous

The reverse mould will be a DRUM. The feature will either be incised in reverse in the drum or run up in positive as a model on the drum, from which a mould will be cast. For this type of running see SPINNING, on curved surfaces and RUNNING CHANGING CURVES, on curved surfaces.

2   A barrel vault in a domed or vaulted ceiling to admit light.

(The term may, however, be applied in plastering to any similar feature. See Fig. 63.)

The reverse mould from which this feature is cast consists of two intersecting drums and is constructed as for VAULTING. The drum for the minor barrel vault is constructed against the drum for the major vault or against the dome mould. Either vault may be of any shape or size; the drums are assembled in their correct projections. The minor

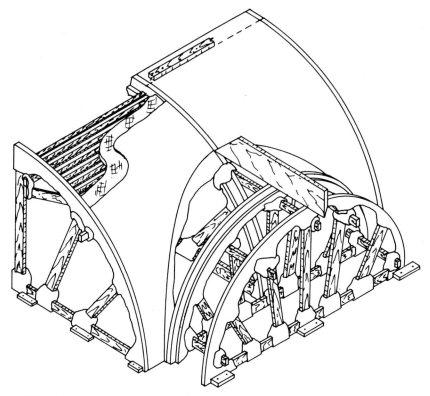

Fig. 63. Lunette (fibrous): constructing a drum for a semi-circular window head to a semi-circular barrel vault

vault is usually small enough for the whole intersection to be cast in one, but where this is not possible, casts are taken from the drum in sections, the groin line being made up with as few pieces as possible.

Casting, fixing and stopping are carried out as for vaulting.

Solid

The work is carried out as for FLOATING TO CURVED BACKGROUNDS. (See Fig. 64.)

MACHINE FINISH, EXTERNAL

See FINISH/RENDERING, EXTERNAL.

MAGNESITE

Magnesium oxychloride flooring. A specialist flooring consisting of the above with fillers – wood flour, sawdust or ground silica.

MAKING GOOD

Fibrous – filling and finishing damage to old work.
Solid – making up to and jointing new work; repairing new and old work.

See also STOPPING; MITRE; PLAINFACE; REPAIR AND RESTORATION.

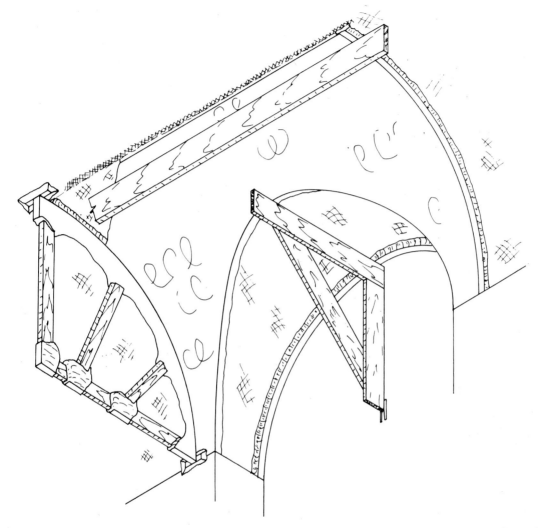

Fig. 64. Lunette (solid): turning the floating to the window head and floating the barrel vault with pressed screeds

MAKING GOOD, EXTERNAL
New work

This will consist of filling in holes caused by construction alterations, scaffolding and repairs to damaged areas.

Preparatory work to all areas

Holes bricked up, joints raked and all edges trimmed to form a regular shape and cut back. Brush dry, dampen and apply a spatterdash coat to assist adhesion.

Backing

Prepare a making-good gauge, cut to allow the full finish thickness. The mix will be either 1 : 3 or 1 : 1 : 6 depending upon the type of finish. See FINISH/RENDERING, EXTERNAL. Apply to background, push well into the edges and rule off with the gauge, key with comb scratcher for all finishes other than Tyrolean and natural aggregate materials.

Finish

As in FINISH/RENDERING, EXTERNAL.

Old work

Remove all loose sections of the existing work, trim to shape and cut back edges, rake joints; dry-brush, dampen and spatterdash. Apply backing as in new work, and finish as in FINISH/RENDERING, EXTERNAL – to match existing work.

Old moulded cement work

Cracks, cut out and brushed. Grout with neat cement and fill with cement and fine washed sand 1 : 2; finish with wood float.

Larger areas, prepare backing as for cracks and proceed as in MITRE, mitres in Portland cement and sand.

MAKING GOOD, INTERNAL

Carrying out repairs to new plastering, making up damaged external angles, plain surfacing and mouldings. Filling in large holes caused by construction alterations, filling in smaller holes that have received pipes and the like and in general tidying up in readiness for the decorator.

Preparatory work to all areas

Clean and make good background, brush off dust and dampen. Clean edges to a regular shape, hack and dampen.

Backing

Prepare a making-good gauge, and mix sufficient sand and Class A plaster, 3 : 1, to fill out the hole to within 2 to 3 mm of the finished surface. Apply and tuck well into the joints and rule off with the making-good gauge. Clean the edges and the plaster surface and key the backing.

Finish

Use either the same plaster that has been applied to the main surface, or a 1 : 1 mix of Class A plaster and lime putty. Apply evenly to the backing and rule off flush to existing work. When firm, trowel hard and clean the edges. Provided the work has been correctly carried out, no joint should be felt between old and new work.

Cracks

These are raked out and cut back, dampened or coated with a weak bonding agent; this will control suction and assist adhesion. Apply a matching plaster, well into the crack, rule off with a joint rule and trowel hard with a dampened laying on trowel or joint rule.

Mouldings

Cracks as above, large areas as in mitres in specified material.

Plasterboard

Cut away the damaged area so that a whole or part replacement board can be fixed securely to wood joists. Finish in Class B board finish as in one-coat finish to plasterboard. For small areas apply a scrim and Class A plaster wad, keep back behind existing finish and finally finish in either neat Class A or Class B board finish.

See also REPAIR AND RESTORATION.

MARBLE, ARTIFICIAL

See SCAGLIOLA.

MASONRY CEMENT

See PLASTICISER.

MASONRY, IMITATION

See FINISH/RENDERING, EXTERNAL.

MECHANICAL PLASTERING

Spray plastering

Several methods are available. Basically, all involve mixing, pumping and spraying. Mixes can vary from cement–lime–sand to pre-mixed gypsum plasters. In some systems the entire operation of plastering from floating to finish is machine operated while in others the finish is obtained by craftsmen using traditional hand tools.

Metal angle beads are generally fixed the day before, and *a clear run is essential*. Every stoppage will usually mean that the delivery pipe must be cleared out. The size of the gang of men required will vary with each appliance, the minimum usually being five. While one sprays, the others will be ruling off, scouring and working on areas too small for spraying. The finishing is sprayed on, ruled flat then scoured with a carpet float, and finally trowelled.

Power floats

These are circular power-driven steel-bladed appliances complete with handlebars and controls. They are used to finish concrete, cement–sand and granite floors. A flat surface should be obtained first and, as soon as it is stiff enough to withstand weight, a finish can be obtained by the power float. In general a dry mix is recommended for this type of work.

Vibrators

Usually in the form of a poker and used to vibrate concrete laid *in situ* or in moulds in order to compact it. Also may be attached to special rules for levelling in large concrete areas. A vibrator is also used to ensure that, when building ferro-cement boats, the keel of the boat is filled solid with cement and sand.

MEMBERS

The surfaces that make up the section of a moulding.

MEMBERS, SETTING

See RUNNING IN SITU.

METAL LATHING

See BACKGROUND.

METOPE

The space between two TRIGLYPHS on a Doric frieze.

MITRE

A mitre is a joint between two mouldings intersecting at any angle other than a straight line.

Fibrous

Mitre cutting

Mouldings on a single surface may be offered on to the setting out, and the mitre line marked on both edges of the moulding. When it is shorter than the length of the saw, the mitre line need not be marked across the contours of the moulding; the saw is merely held vertically, and sighted as a billiard cue through the two marks. With right-angle mitres a 45° set square may be used to mark the mitre line on the moulding.

Mouldings not resting on any one surface need to be held in projection relative to their position when fixed, in order that the mitre may be cut plumb. With cornice this is achieved by holding the cornice upside down with the ceiling strike-off flat on the working surface. A saw held plumb will cut a plumb mitre. When setting out right-angled mitres in cornices, the wall length is first marked square across the cast and the projection of the moulding is then marked on the ceiling line in front of the square line, for an external mitre, and behind it for an internal mitre.

When such mouldings intersect at angles other than 90°, the distance that the ceiling lines intersect in front or behind the square line is marked on the ceiling line of the cast. The mitre is cut either by holding the saw plumb and sighting through the marks or, in large mitres, by marking out the mitre line over the section with a square.

Mitring

Where it is possible to get at the back of casts fixed on site, wads to join the two casts together should be fixed behind the mitre. Otherwise, a sufficient gap should be left in the mitre to allow a substantial wad to be poked in from the front. Small mitres should have their stopping reinforced with strings of canvas, if not with wads. Unreinforced stopping to joints and mitres between casts will probably crack. In shop work, mitres worked in on moulds should be cut undercut and similarly reinforced, whereas mitres in models off which a single mould is to be cast need no reinforcing. When casts are mitred, the mitre should be checked to ensure that all members pick up. Larger casts often twist between casting and drying out so that some members may pick up, while others do not. To adjust such a mitre, parts of the section may be fractured and bent, or cut and bent. Dry mouldings are damped to prevent excessive suction. The plaster is gauged to a creamy consistency and the canvas put in to soak. The soaked canvas is pushed into the gap as the plaster begins to pick up and the joint rule passed over both sides of the mitre to ensure that all canvas is down.

The technique of working the plaster is to apply it with a trowel-end small tool to follow the members slightly full in order that it may be cut off with a joint rule as it stiffens. Stiffening will be brought about by a small amount of remaining suction and the set of the plaster – too much suction will cause the plaster to work up woolly. The joint rule should be held on the moulding strictly in line with the members and perpendicular to the contours. To avoid tearing lumps of plaster from the mitre, the joint rule is worked into each member and arris in the section, first from one edge and then from the other. This is done all down one side of the mitre and then down the other.

The finish is achieved by repeatedly applying plaster during its setting period and cutting it off. The cleanest, sharpest finish is achieved by cutting plaster in an advanced cheesy stage with clean steel. Except when an 'antique' finish is required, a water brush should not be used. The cheesy plaster should not be knocked back as a whole but sufficient for each application scraped like butter from the surface, leaving the rest undisturbed for further use. Once the plaster has been re-tempered it will set quickly and cannot be re-tempered. Over-re-tempering of plaster in a too advanced stage of its set will kill it. Generally it is better to finish a mitre with one gauge rather than rough out the section first. Mitres that cannot be completed in one gauge can be completed in sections. However, when members need excessive building up to make them intersect this work may be done with a preliminary gauge.

Soaking the plaster is an alternative to stirring it. To gauge in this way, the plaster is merely sprinkled into the water and left to set. The advantages of this method are that: (i) Plaster of varying stiffness can be selected from different parts of the container, (ii) Because it sets more slowly, it will allow larger mitres to be completed in one gauge.

Internal mitring can be made easier if a scribed effect is produced by allowing one moulding to carry through, a few millimetres beyond the intersection. This requires the making up of only one moulding into the other, providing a firm surface against which the point of the joint rule can be worked.

Scribe

The shaping of the end of a moulding so that it will fit the contours of another moulding as it butts against it to form an internal mitre – as opposed to cutting both mouldings across a mitre line. The mitre line needs to be marked accurately across the section, either by setting out or by cutting an internal mitre. The scribe is then formed by trimming slightly slack of the cut with a saw held at the angle at which the other moulding will meet it. The end of the cast will then be trimmed to a shape that will fit the face of the other moulding. See also previous paragraph and STOP, SLIDING.

Curved mitres

Where a curved moulding meets a straight one, it is easier to allow the curved moulding to run through on internal and external mitres, as previously described. This requires no making up of the curved moulding. The internal portions are as a scribe while the external mitre is carved in the curved moulding by a joint rule worked on the straight, leaving a joint to be filled in just short of the mitre line. Where two curved mouldings intersect, however, at least one will have to be made up, the best way being to cut and file curved joint rules from busk, saw blades, joint rules, etc. Both concave and convex may be needed to fit the curves of the moulding.

Mitring enriched mouldings

1   The cut can be made to balance the enrichment about the mitre line, and the enrichment stopped in by modelling and carving.
2   The cut is made as for (1) above, and the intersection covered with a MITRE LEAF. The leaf may also conceal any imbalance in the enrichment.
3   The enrichment is left off the end of the cast or cut off on site and its bed made up. The bed is mitred along with the rest and enrichment, cast separately, planted in the bed to the mitre and finished as either (1) or (2).

Solid

Forming in putty and plaster

Solid-formed mitres, mainly because of the size of the running mould, are usually much wider than mitres formed by fibrous plastering. The minimum length of each side must be equal to half the length of the mould slipper. Wherever possible the mitres should be roughed out well behind the finished line while the running operation is proceeding. A good craftsman can do this freehand with the toe of a gauging trowel; it should be well keyed.

After the running rules have been struck and any surplus material removed from the mitre edges, the work can begin. The gauge should be the same as for running, at least 50% putty, 50% plaster, and this should be well mixed on the board. Do not use immediately as the mix will probably be too wet, either wait till it gets cheesy or take a little away from the main mix and stiffen with a little extra plaster.

Apply this material to a roughly formed moulding shape to the top members of the cornice, either side of the mitre. Some plasterers will use a trowel-end small tool for this, others prefer a small gauging trowel. Then using a joint rule that should be at least half as long again as the mitre, and holding it in both hands – the right hand over the back edge, the left hand under, for right-handed people – keep the joint rule parallel to the members, and the heel firmly but gently on the finished work, and draw downwards in a stroking action. Do this to both sides, checking that the mouldings intersect perfectly at the mitre. When these members are shaped out, lay material on the next two and repeat the whole operation till the mitre is complete.

At this stage a fresh gauge of rather wetter putty and plaster should be mixed and applied to the whole mitre. A splash of water into the angle before application may be necessary to counteract the suction. Repeat the operation with the joint rule, and the result should be a perfectly formed clean mitre, with good intersection at the angle. Should

the plaster at any time get too hard to remove with the stroking action then use the edge of the joint rule to cut it – that is why it is ground or bevelled. Where possible it is best to try to avoid this, as sometimes the extra pressure involved will result in damage to the finished cornice. Always work away from arrises.

When working on smaller mitres, such as coves and the like, a good craftsman will probably put two or more in at one time. However, the basic instructions still apply. Do not finish with a brush.

Forming in Keene's cement

Keene's plaster is not currently commercially available but this entry may be of use in restoration work. Because of the slowness of set in Keene's cement it should be possible for the plasterer to rough out or rough shape most of the mitres in the room and then return to finish them. At all times one should resist the temptation to dope the Keene's with ordinary plaster. Each mitre should be cored out, either during the running or afterwards, and then shaped out as with putty and plaster. For the finishing it may be necessary to sift a little Keene's as in running and this will produce a finer smoother finish.

The forming of the mitres in this material is normally carried out on the day after the running is completed. Therefore little damage will be done to the cornice. However, the same rules apply to the use of the joint rule. Work parallel to all members and stroke into position rather than digging or forcing; always work away from the arrises.

Mitres in Portland cement and sand

It is essential that mitres in this material are worked upon whilst the cornice is being run. Internal mitres should be built out to the same degree as the cornice and ruled in so as to intersect perfectly with purpose-made wooden joint rules. External mitres should be run up in position wherever possible, using either the backed rule or notched rule method.

Provided all of these jobs have been carried out in the correct sequence the mitres will be finished at the same time as the cornice, with small and shaped wooden floats. Breaks and returns should also proceed at the same time as the running. See BREAKS AND RETURNS.

Mitres in solid run mouldings: summary

A good mitre in many instances is the hallmark of a good craftsman. The tools required to carry out this job effectively are many and varied – several metal small tools of varying shapes and sizes, joint rules from 100 to 450 mm in length, several good small tool brushes and at least one good small gauging trowel. The joint rule cutting edge must be kept straight and well ground.

For Portland cement work the craftsman will need wooden joint rules; these are usually made from old and broken feather-edge rules and the like, and again the cutting edge must be kept straight. He will also need many small wood floats of various shapes and sizes.

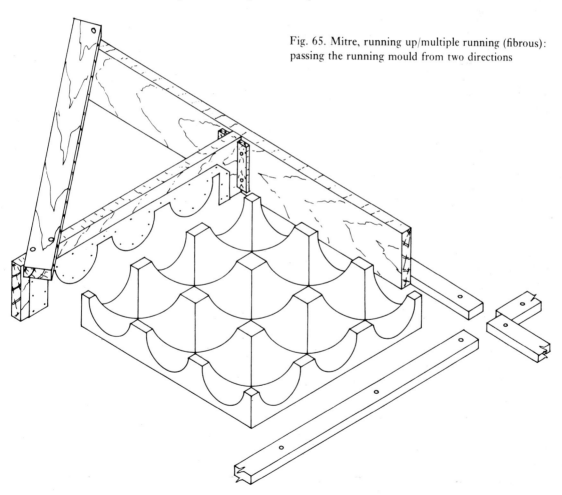

Fig. 65. Mitre, running up/multiple running (fibrous): passing the running mould from two directions

Though in most cases it is common practice to pass a wet tool brush over the finished mitre this is not entirely necessary. Things to look for are the continuation of line, the intersection and the finish. All of these should be possible to the skilled plasterer without the help of a wet brush. Brushes are required for damping in and, in certain cases, applying small quantities of wet material to awkward places. The final essential for good mitring is good eyesight.

MITRE LEAF

A leaf, cast in plaster and used to cover the intersection of a line of enrichment at a mitre – external or internal.

MITRE, RUNNING UP/MULTIPLE RUNNING

External mitres may be run up by running intersecting mouldings simultaneously, one instance being a cornice capping a free-standing pier, using notched rules.

One running mould may be used to form a feature by running over it from different directions, the running rules being suitably notched to allow the passage of the slipper (Fig. 65).

Figure 66 illustrates a feature formed by different running moulds being worked as in Fig. 65.

Features such as a multi-sided baluster may be run up by repeatedly moving the spindle round, fixing it in predetermined positions, and passing a running mould over the top of the box. (See Fig. 126.) The spindle is formed in the usual way and the

feature cored out with a MUFFLE on the running mould. See TURNING box.

MITRE STOP

See STOP, SLIDING.

MODEL

Any object from which a mould is taken (originally known as a solid or a moulding piece). It may be made up new by a plasterer or modeller/carver, cast from a mould (usually with a view to storing it as a stock model), or be a piece of old plasterwork taken down for reproduction. See also under specific item.

MODEL MAKING

See under specific item and especially ENRICHMENT, planting.

MODILLION-BLOCK CORNICE

See CORNICE, MODILLION-BLOCK.

MODILLION BLOCKS

Rectangular blocks, enriched or plain moulded, appearing as a line of enrichment in a cornice or

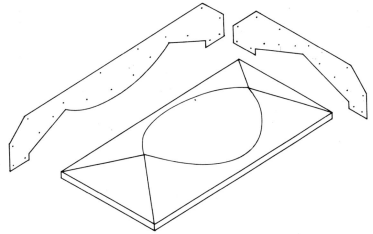

Fig. 66. Mitre, running up/multiple running (fibrous): a tile run up with two different running moulds

allied feature. Unlike dentil blocks, they may have a moulding breaking round them and may alternate with PATERAS. Where they appear in a feature supported by columns they are spaced so that there is a block centrally over each column. For making up the model see BRACKETS.

MODILLION BLOCKS, RAKING

See RAKING.

MONOLITHIC FLOOR

See FLOOR LAYING AND ROOF SCREEDING.

MORTAR

A mixture of sand, lime and water. In plastering it will usually mean raw coarse stuff with or without HAIR. Once cement or other materials that will cause it to set are added the word 'gauged' is used before mortar or coarse stuff.

MOULD OIL

A water-soluble vegetable oil used as a release agent when casting concrete.

MOULD, REVERSE

A reverse mould is an object from which a cast is made. It may be formed directly by running it with a running mould or ruling it in from such surfaces as grounds, rules and templates. Alternatively, it may be an impression taken off a model and is made by casting from the model in a suitable moulding material, the most common of which are plaster, hot-melt POLYVINYL CHLORIDE; COLD POURS; GLASS FIBRE; WAX; GELATINE and CLAY.

Plaster moulds

The advantage of using a plaster mould, whether it is run or cast, curved or straight, is that the lines are rigid and therefore will remain constant, producing good and undistorted lines on all casts. This is especially beneficial with lengths of mouldings where the members form many closely spaced lines. Plaster moulds are, however, at a disadvantage when it comes to coping with undercut as the material is non-pliable. This problem is dealt with by the use of waste moulds or piece moulds.

Plaster moulds – run

A mould formed directly by running with a running mould. See CORNICE; BEAM CASE; NICHE.

A run mould, the section of which contains an undercut or vertical parallel sides, will require that part of the mould to be loose so that it can be moved aside or come out with the cast. This may be achieved in one of two ways: the loose piece is formed on a bed in the reverse mould as the section is run up and is termed a run loose-piece mould. Alternatively parts of the section may be run separately, or even with different running moulds, and brought together so that when assembled they make up the required section; this is termed a run slip-piece mould.

Plaster mould – run loose-piece

Figure 67 (a) shows an undercut plaster reverse mould run with a running mould in such a way that the undercut member or members along its length are run up with the rest of the section but are loose so that they may come away with the cast.

The metal plate that will form the bed for the loose-piece is fixed in position over the original profile. The muffles are then cut and fixed over the rest of the section, care being taken to form the separating arrises by stopping them about 5 mm short of the plate. The core can then be run off. The bed for the loose-piece is shellacked, french chalked and greased, while the rest of the core is well keyed to receive the finish. The plate and muffles are removed and the running completed.

In the finished section the loose-piece, produced with the same plaster as the finish on the rest of the mould, is separated from it by the arrises that were formed on the core against the original profile between the plate and the muffles. The core will expand on setting and the profile will foul the two separating arrises. This can be counteracted by removing the plate and muffles immediately a satisfactory finish is obtained and by continuing to pass the running mould over the section, cutting off the expansion until it is complete.

The alternative is to cut the expansion from the fully set arrises slowly by lightly and repeatedly passing the running mould over. Such a loose-piece section, if the length of a conventional cornice mould, is likely to break during use. It may either be reinforced with a metal bar and canvas or cut into shorter sections. This latter method will produce small seams which, however, are easily cleaned from the face of the cast.

In general, when determining the section that is to be loose, the design points to observe are:

1 The piece should be retained in its bed by gravity.
2 The undercut should be efficiently dealt with by the loose-piece, i.e. neither should undercut be left in the body of the mould nor the piece (having removed too much of the undercut) itself be irremovable from the cast.
3 The joints are made at convenient places relative to the members in order that the seams may be easily cleaned from the cast.
4 As with all piece moulding, no acute quirks should be contained within the bed of the loose-piece as they will collect dirt and prove difficult to clean out, thus preventing proper seating of the loose-piece.

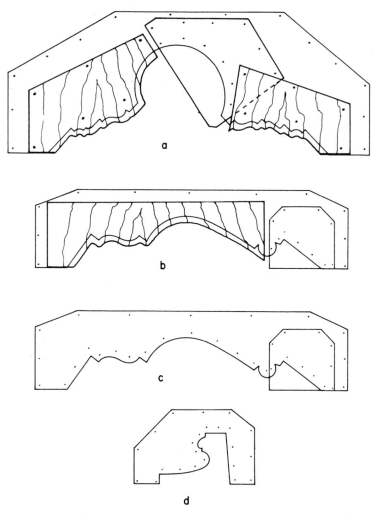

Fig. 67. Mould, reverse (fibrous): (a) run loose-piece, (b–d) run slip-piece

Plaster mould – run slip-piece

A run piece mould where the loose-piece slides across the face of the bench or drum, rather than sitting in a bed contained by the mould. It may be produced by the method for run loose-pieces (Fig. 67 (b)) or run in two or more separate parts and put together. See also BEAM CASES. Figure 67 (c) shows part of a section masked off with a metal plate and the remainder run loose. This may then be slid to one side. To ensure accuracy, a second plate is fixed against the first and over the remainder of the section. The first plate is then removed. The second part of the section may now be fixed down to the bench. The larger of the two sections is less likely to break and so is made the loose slip whenever practicable. The mould is set up for casting by cleating the loose run against the fixed section with the aid of folding wedges. The handrail in Fig. 67 (d) shows how, where the section is symmetrical about the undercut, both sides of the mould may be produced by the same profile, the first run loose and the second fixed.

If one cast only is required, a nail projecting from the stock may be used to separate the reverse mould into two sections along the undercut. The nail is left in position until the section is fully built up with the first gauge. When it is seen that the mould will be finished by just two or three more runs, the nail may be removed so that the groove will be filled with the finishing plaster. This will not bond the two sections together and the mould can be split apart, thus releasing the cast.

Plaster mould – cast

The reasons for choosing to produce a plaster mould by casting it from a model, involving an extra operation, are:

1   When, because of its shape, producing a section in its positive rather than negative form may be an easier task.

2   When the different method employed to produce a positive section will result in a more geometrically accurate shape, for instance, with a mould for a niche with an elliptical head and shaft.

3   Where a mould is made up of many run mouldings joined and mitred together (with or without ruled-in panels) it will break up during repeated castings. It is therefore the model that is laid down in such a way, and the mould is cast off it as one reinforced unit.

4   The face of a run moulding is formed with knocked-back and thus weakened plaster while its arrises are filled and polished with a softer gauge. On the other hand, the face of a cast is formed with thicker, fully set plaster and made even harder by the effect of size. Therefore, if many casts are to be taken from a mould, it may be preferred that the mould be cast rather than run. Special hard plasters may be used (or even GLASS FIBRE). In this instance, a model is run off with moulding grounds and a mould cast from it containing reinforcement as for a case on a run-case mould.

5   A mould which has been cast is portable and may therefore be stored for use as a stock mould.

6   When moulding in the round or when undercut is present on a model, a plaster piece mould or waste mould may be chosen for its good lines and rigidity, thereby avoiding distortion.

Plaster mould – cast – piece moulding

A mould made of separate pieces. When small in number the pieces are clamped or bound together (see Baluster (Fig. 11)); larger numbers are held together by fitting into a cast fibrous plaster back. (See Fig. 68.)

The model is first sealed, if necessary, in the usual way for fibrous plaster casting. Sheets of clay are then beaten to an even thickness on a clay board. The clay is cut into strips and used as walls between which the plaster pieces will be formed. The size and shape of each piece is governed by the undercut on the model. Each piece is splayed so that it will draw from the back and from the piece next to it. Usually the pieces are held in position by a saddle or back which itself may consist of one or more pieces. If the shapes of the pieces do not make them self-locating one with another or with the back, joggles may be introduced.

The area within the walls is greased if necessary and each piece is cast in turn, using the usual technique – first brushing the plaster well into the detail on the model and splashing on a thickness with a brush. The piece is then built up to the required thickness, usually in the region of 30 mm, with a small tool or gauging trowel and worked

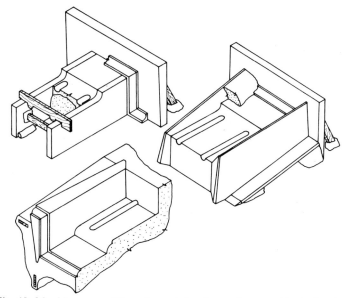

Fig. 68. Mould, reverse (fibrous): model and cast piece mould of a bracket

smooth as it sets. The clay walls are removed, the piece trimmed if necessary, shellacked, french chalked and greased so that it is ready for the erection of clay walls and casting of the next piece. On no account should any piece be removed from the model – it will never fit back properly. When all the pieces have been formed a back is cast over them and wooded with braces in such a way that it will rest flat on the bench while casts are being made from the piece mould.

When designing the pieces of a piece mould the points to observe are:

1   No piece should be locked immovably either to the model by undercut or to adjacent pieces by incorrect splaying of its sides.
2   The general surface formed by the backs of the pieces should be smooth and splayed in such a way that it is free from undercut, permitting the eventual withdrawal of the fibrous plaster back to be cast over it.
3   Sharp quirks to locate the pieces within the back should be avoided as they will easily fill with dirt

and prove difficult to clean out, thus preventing the pieces from fitting properly into the back.
4   No canvas reinforcement is used in the pieces as this would cause each piece to twist when the plaster expands on setting. In moulds where no back is used and each piece is reinforced as an individual cast, as with balusters, a special non-expansion plaster may be used. The use of this plaster will also eliminate the possibility of a step in the pick-up on a cast, caused by one piece of the mould having expanded more than the other. These plasters are, however, up to four times as expensive as normal casting plasters.

For further details see under specific model to be piece moulded.

Plaster mould – cast – waste moulding

The plasterer, having many different moulds to choose from, usually limits the use of waste moulds to large models modelled in clay and requiring only one cast: large because the use of an appropriate

PVC mould would be troublesome and costly; clay because this material can be scraped from an undercut mould; and one cast only because both model and mould are destroyed in producing the cast.

In sculpture, waste moulding is widely used though pliable moulding materials (in the form of case moulds and back-and-front moulds) are made use of on small models. Here it is far more convenient to use waste moulds in the form of piece moulds when moulding undercut models that have been sculpted from rigid materials.

A clay model will need no preparation; glass fibre, metal or other dense materials will merely need greasing. Although wood and other porous materials are best sealed with shellac and greased, the traditional treatment of plaster is to coat it with a soft-soap lather or a clay wash.

For basic casting techniques see CASTING IN FIBROUS PLASTER, general method. The firstings of the mould is applied to the face of the model more thickly than normal – somewhere in the region of 10 mm – care being taken to get the plaster into all the detail without damaging the face of the model with the brush. The later chipping of the mould from the cast is greatly helped if the firstings contains a strong-coloured dye, usually in the form of a powder added to the gauging water. Chipping off the back of the mould is also made easier by giving the firstings a smooth rather than a rough back. The undercut nature of the mould will lock the firstings to the back. When set, the back of the firstings is coated with a release agent – traditionally a soft-soap lather or a clay wash. The rest of the mould is then cast over the firstings with normal casting plaster, firstly brushing and splashing on a coat of seconds and then proceeding in the normal way.

The reinforcement of a mould is kept to a minimum as its presence makes the operation of chipping the mould from the cast more difficult. If the canvas is kept to one layer it will assist in the removal of the mould by allowing it to be stripped off the firstings in layers. Large casts are reinforced by wadding on struts and braces.

If a model containing undercut is made of a rigid material, such as glass fibre, a cement mix, wood, stone, etc., then the waste mould will have to be cast in separate pieces, each piece being formed so that it will 'draw' from the undercut. Strictly this would be classified as a piece mould rather than a waste mould. Where the model is of clay or similar material, undercuts can be moulded. When the mould is set the clay is scraped from the inside and the mould may be washed out with water. This can be done in such a way that the required amount of clay wash to act as a release agent is left on the face of the mould, assuming that clay has been chosen rather than soft soap.

When moulding clay models in the round, however, a waste mould of more than one piece will have to be used. This is because the mould will have to be split in order to remove the model and to fill in the cast. The cast can be joined by assembling the mould as the final part of the casting operation, or the mould removed from the pieces of cast and these pieces joined together and made good afterwards.

The casting is carried out in the normal way according to the type of cast being made.

When the cast is fully set the mould is chipped from it with a mallet and wood chisel. Any braces are first cut off and the plaster back removed from the coloured firstings. This operation is usually comparatively quick and easy, the layer of coloured plaster warning the operator of his approach to the cast. The coloured layer is then very carefully chipped from the face of the cast. Any chips and chisel marks on the cast are later made good.

Plaster mould – formed

These moulds are neither run nor cast but are made up of surfaces ruled in and finished from ribs or grounds. This method is used when they are basically composed of large plain surfaces that would

154

prove more difficult to produce by running. Although surfaces contained by the mould may not be curved, they are usually known as DRUMS. They may be used (a) to produce plain, flat surfaces (see DOME; VAULTING); (b) to shape PVC moulds, placed over them (see MOULD, REVERSE, flood); (c) as a base to receive integral moulding and insertions produced by running, either directly over the drum or run/spun elsewhere and placed upon it. See BEAM CASE; NICHE, curved backgrounds; MOULD, REVERSE, run-case – insertion.

Case moulds

The face of a case mould is a film of pliable moulding material of even thickness, supported by a fibrous plaster case. Case moulds differ from flood and skin moulds in that the case is made first and the pliable moulding material poured into the void between the case and the model. To produce this type of mould a core is formed over the model to the required shape and thickness in a material that will subsequently be replaced by a pliable moulding material. This is achieved by casting a plaster case over the core, removing the case and the core, replacing the case and flooding the void created by the absence of the core with the pliable moulding material.

The advantages of case moulds over flood and skin moulds are:

1 Large and deeply contoured models may be covered efficiently to a controlled thickness with a minimum of material.

2 This method avoids large pockets of pliable moulding material at the bottom of contours. Such volumes of pliable material, shrinking on solidifying, will contract markedly and distort the face of the mould.

3 Lips and joggles can be formed to lock the material into the case, both round the strike-offs and on vertical surfaces where it would sag and bag under its own weight.

Two types of case mould are available: clay-case moulds and run-case moulds. With the former, clay is used to make the core. In general, they are used for models and enrichment in the round. Run-case moulds are formed in plaster by running with a running mould. Their use is, therefore, confined to lengths of enriched mouldings, curved or straight, the advantage over clay being that the smooth and parallel case will allow the strip of pliable moulding material to be stretched or shrunk along its length, so allowing the enrichment to pick up in joins or balance in mitres. In addition, members can be formed along its length to hold the pliable material to a good line and may be placed in any similar bed in any reverse mould made to any shape.

Case mould – clay-case and back-and-front mould

Although clay-case moulds can, of course, be used to reproduce lengths of enriched mouldings, when the preparation of the clay is taken into account it will be seen that it is just as quick to run a plaster core. This will have the advantage of a run case, producing a mould that can be stretched and shrunk, that will contain lips and controlling arrises, and that may be placed into other cases run to any shape.

The use of clay cases is, therefore, usually confined to models in the round. If the contour of the enriched model contains no major undercut, features in the half-round, such as trusses, are normally moulded with a one-piece mould. Models containing major undercut, such as the volutes on an Ionic pilaster cap, may have the moulding compound in one piece but the case will have to contain loose-piece cheeks that will draw from the back of the mould in order that the compound may be peeled off the model. With such models the case is cast from the core as a piece mould.

When a model is completely in the round or contains such undercut that the compound cannot be peeled off the cast, the whole mould is made in pieces and is termed a back-and-front mould. The clay core is formed and clay walls erected to cast the case in pieces as for piece moulds. The size and shape of each piece is determined by the undercut. All pieces of the case are cast before any is poured. Using PVC or cold pour, one piece at a time is then removed, along with the core underneath it, the case replaced and poured, this process being repeated with each piece in turn. Lips may be formed on the core, against the walls, so that each piece of the compound is retained by its own case. When gelatine is used, however, as many of the pieces as possible are poured together and the jelly is cut on removal from the model with a jelly knife. This special knife has a groove in the blade that forms a male and female locating groove on adjacent pieces of the jelly.

It is seldom possible to cast from such moulds when they are assembled and the usual procedure is to fill each piece separately. To join the casts together they may be cast simultaneously and bedded together with a little surplus plaster on their strike-offs as the final casting operation, and the mould removed when the cast is set. Alternatively, each piece is removed individually from its mould when set, then all are bedded together. With either process the joints should be wadded up from the inside where possible. All seams are then cleaned up and made good.

Clay-case moulding – basic method

All models likely to move during the moulding process are fixed to the bench. Sheets of paper are dipped in water and laid over the model to protect it from the clay core. Any minor undercut affecting the removal of the case is blanked off by filling it with clay. Sheets of clay, beaten to an even thickness – 6 mm for cold pour, 15 mm for hot melts – are then laid over the model, joining each piece invisibly to the next. A continuous lip is formed all round the perimeter to lock the edge of the moulding compound into the plaster case. Any lips and ledges within the mould are also formed at this stage (Fig. 69). When hot melts are used the pouring funnels are always located in the lowest part of the case in order that the compound will rise uniformly round the model, pushing out the air and avoiding the formation of seams. A funnel placed at the highest point will cause the compound poured through it to run down over the model like syrup over a suet pudding, forming seams at every run. Where it is essential that a funnel is placed at bench level, provision is made in the clay to extend the core so that this may be done.

The core may be smoothed as a whole by rubbing it with a wet piece of canvas. A smooth, silky sheen may be produced by then rubbing with a paraffin rag. The paraffin will also act as a release agent. The pour holes may either be cut in the plaster cast or formed by striking off when casting from the top of clay plugs on the case. The fibrous plaster case is cast and reinforced to suit the type of case mould being made, the final braces being positioned in such a way that the mould rests flat on the bench at a convenient angle while casts are being made from it.

When the case rests on a flat moulding ground and is not automatically located, a pencil line must be drawn round the case before it is removed from the model so that it may, without damaging the model, be placed back in the correct position for pouring. The clay case may now be removed, along with the clay core and paper, and the model seasoned according to the moulding compound being used. Pour holes, if not already formed in the case, are cut in the lowest places. With hot melts, these should be some 70 mm in diameter in order that the case does not chill the compound during pouring. 6 mm air holes are drilled in the lips and in any hump that will cause an air pocket to be formed in it by the level of the rising compound.

Any frayed canvas is cut or burnt from the edge

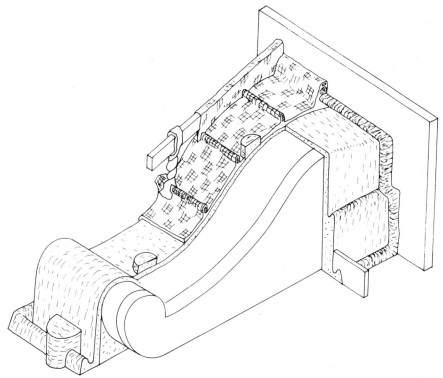

Fig. 69. Mould, reverse (fibrous): clay-case mould cut-away showing model, paper, clay or pliable material and cast case

of the holes, the case dusted free of all loose particles, seasoned according to the moulding compound being used, and carefully placed back over the model. In order to prevent loose particles falling through the pour holes on to the model, it is advisable to erect the funnels next and cover them with paper. A seal is made between the funnel and the case with an encircling ring of clay which will adhere more effectively to the case if the plaster is shellacked. The funnel may be further secured by a wad placed over the clay and joining the funnel to the case.

To effect easy removal of the wad, minimum contact may be made between it and the case. The case is secured over the model by weights, struts or wads. The wads are hooked over the edge of the case and,

so that they do not adhere to it, may be placed on strips of paper. When moulding in the round, both ends of the wad are treated in this way. When the case is to be fixed to the bench the wad is poked round two nails toshed into the bench against the side of the case. The joints between the case and the model or bench will be tight enough to prevent hot melts from leaking out. Cold pours, however, are not congealed by the cold surfaces and will creep out, making necessary the sealing of such joints with plaster.

With hot melts, pouring should be continuous – a pause in the flow may cause a seam to form in the face of the mould and delays may cause the material to congeal and prevent further pouring. Pouring should begin at one funnel, and the compound be allowed to flow from the air holes for a little while before they are plugged with clay. The material flowing first from each hole will contain entrained air and, if allowed to remain, will form a soft, spongy, honeycomb pocket in the back of the mould. Pouring should not begin at the next funnel until the material has reached it and started to rise up it, otherwise an air pocket will be formed between the two funnels unless there chances to be a convenient air hole. Honeycombed material and air pockets will cause the face of the finished mould to sag, thus forming a bump on the cast which, on enrichment, will be virtually impossible to remove. When the mould has been completely flooded the funnels should be topped up to maintain the pressure and act as reservoirs for replacement of shrinking hot melts.

When the compound has set, the funnels may be removed, the compound cut flush with the case and the case released by removing the wads. The case may now be lifted clear of the model, turned over and placed on the bench. Because of the enrichment, the model is usually more undercut than the case and will retain the moulding compound. If, however, it comes away with the case it must be removed. The case must now be treated – if this has not already been done – to house the compound. Nipples formed in the air holes will not sit back in these holes and will have to be removed. The strike-offs and sides of the case need to be seasoned to suit the material to be cast from the mould. Where necessary, the pliable material is also treated for casting.

Case mould – run-case

Two methods exist for producing the case: it may either be cast in fibrous plaster from a run core, or run with a running mould as a run cast.

CAST FROM A RUN CORE

A model that can be laid down on the bench with the bench as the moulding ground need not have the core run over it. It may, in fact, be run separately on another bench with a separate running mould or an adaptation of the original running mould. If, however, the moulding grounds are not in the same plane and have to run up as an integral part of the model, the operation will be simplified by forming the core and case over the model, (Fig. 70(a)), rather than running up identical moulding grounds elsewhere.

If the core is to be formed over the model, time will be saved by the running mould being initially cut and horsed with the metal profile to run off the core. The profile for the model is fixed in position over the one for the core, backed up with plaster if necessary and the model run off. On removal of the model's profile the running mould is automatically ready to run the core over the model from the same running rule. An alternative is to horse a second running mould with rebates in order that it may be run off over the model by bearing on its moulding grounds.

The profile to form the model is cut and filed to the section of the moulding containing beds to receive each line of enrichment. When the model has been run off it is dressed by planting the enrichment in the beds. When reproducing enriched mouldings it is important that the enrichment picks up when the casts are placed end to end. For dressing the model see MODEL MAKING.

The moulding grounds need to be seasoned as they will have both the sides of the core and the case formed against them. They are sealed with SHELLAC. The model is protected with a layer of paper from the plaster that will form the core. At this stage, no undercut must be presented to the core by the model. All undercut is, therefore, filled with rolled paper before sheets of newspaper or plaster bags are cut to size, dipped in water and laid over the model.

These should be applied from the end of the run back to the beginning, overlapping them by about 70 mm. The overlap produced will be from right to

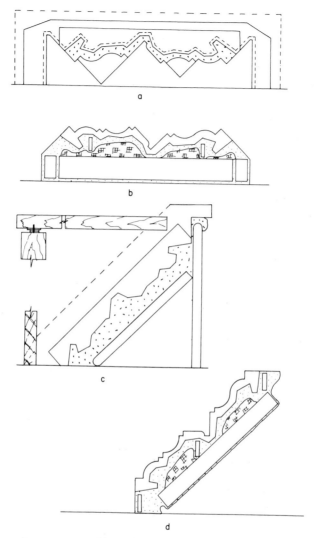

Fig. 70. Mould, reverse (fibrous): run-case mould, straight – (a) the run core on the model, (b) the pliable mould in the cast case. Run-case mould, curved – (c) spinning the core, (d) the pliable mould in the cast case

left so that plaster being moved from right to left by the normal action of the running mould will not be forced under the joint on to the model. The covering paper is carried on to the moulding ground down both sides of the model. A plaster-tight join is made by sticking the edges down with grease or by tacking them down with a lath, or both (see Fig. 71). As a precaution against the wet paper tearing and letting the plaster on to the model during the early stages of running the core, creamy plaster can be pasted over the paper and allowed to set.

The metal profile to run the core is cut and filed

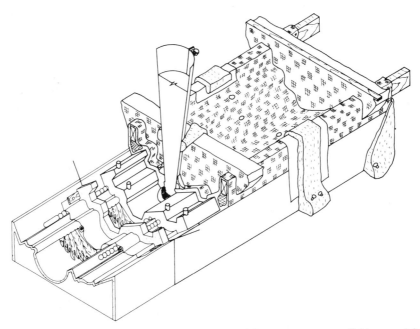

Fig. 71. Mould, reverse (fibrous): run-case cut-away showing model, paper, run core, pliable material and cast case

to follow the contours of the model (15 mm away for PVC and gelatine, and 6 mm away for cold pour) in a series of flats and arrises with a lip on either edge. The pliable moulding material will be locked into the case by the lips. The core is run off in the usual way. (Although sawdust, vermiculite, or slosh from the tank are sometimes gauged with the plaster to form the core, the best results are obtained using neat plaster.) It is then shellacked, french chalked and greased and the plaster case cast from it.

The case has integral 25 × 50 mm battens on ropes forming the edges running down both sides with 25 × 50 mm cross-brackets, covered by curtain wads or laps, spaced every 300 mm along its length. When set, the case is removed from the model. Usually the core will come away with the case as the two side lips will form an undercut. To avoid distortion of the case it should sit evenly on the bench. When the back of the case needs to be straight, a line of plaster can be placed on each cross-bracket

and ruled in with a straight-edge rule off a 6 mm lath placed on the bracket at either end. If, however, it is desired that the case fit the bench it is to rest on during casting, whatever its shape, the surface of the bench is greased.

When the case is fully set a bowl of creamy plaster is gauged and, as it picks up, is placed in a line along each bracket. While the plaster is still soft the case is lifted, turned over, and placed on the greased bench. Each part of every cross-bracket will now fit the bench snugly.

Cracking the core down the middle with a wood chisel and mallet should release it from the case. The core is completely removed and any damage made good. Air holes and pour holes are now drilled and cut in the case. With hot-melt compounds the pour holes must be large enough (some 70 mm in diameter) to allow the material to flow freely. The cooling of the compound by the case will cause congealing if the hole is too small. The cooling factor is not so crucial with gelatine, which is still fluid when it is poured at blood heat, but large holes are still required because of its thick consistency. Pour holes cut alongside cross-brackets will allow funnels to be secured to the brackets as an extra safeguard against knocking over when they are full of molten material. With hot melts, the distance between the funnels will be determined by the amount of material that can flow through them. This amount must be sufficient to cover completely that part of the model served by each funnel; thus, the wider the section the closer the spacing.

For instance, a section of 500 mm girth poured in PVC 15 mm thick will require funnels spaced about 600 mm apart down the centre of the section. The holes may best be cut by drilling a hole to take a pad-saw blade with which a disc may be cut out. Frayed canvas round the edge of the hole may be burnt off or cut away with a sharp knife. 6 mm diameter air holes are drilled every 400 mm or so apart, along every arris in the case that will have an air pocket formed in it by the rising level of the moulding liquid.

All paper, etc., is removed from the model, and the model and case treated prior to receiving the moulding compound. The case should be dusted over to remove all loose particles that may fall on to the model as it is lifted back on. With a model that has integral moulding grounds the case will be automatically positioned, having been cast from them. Cases cast from cores run separately, however, will require lines on the bench to position them centrally over the model. Care should be taken in both instances to lower them slowly at the correct angle in order that the case does not bang against the model.

To prevent foreign matter falling down the pour holes, it is advisable to erect the funnels before the case is fixed down. Each funnel is placed upright over its hole in the case, and a seal effected with a ring of clay. A piece of paper or other form of cover may now be placed over the top of the funnel. The funnel may be fixed more securely by an encircling wad placed over the clay. The area of contact between wad and case can be kept to a minimum so that it may be removed easily by pulling.

The case now needs to be fixed securely over the model. Small cases can be effectively strutted from available beams or ceilings, or weighted down by sacks of plaster, etc., but normally cases are fixed by wads. Although wads may be hooked over the back of the case, a fixing will have to be provided for them on the model or bench. This is usually effected by toshing a pair of nails into the bench, tight against the model for each wad. Wads are more easily removed from the back of the model, causing no damage through their having adhered to it, if a piece of paper is first placed where the wad is to lie. The wad will hook over the edge of the case, thus clamping it to the model. Care should be taken not to block any air holes during the wadding operation.

The ends of the case must now be blocked off with clay and/or wads to prevent the moulding material from flowing out. Air holes will also have to be provided at the extreme ends of the arrises. These may take the form of V-cuts in the end of the case, or holes drilled in the end stops. The joints between the case and the model or bench will be tight enough to prevent hot melts from leaking out. Cold pours, however, are not congealed by the cold surfaces and will creep out, making necessary the sealing of such joints with plaster.

With hot melts, pouring should be continuous – a pause in the flow may cause a seam to form in the face of the mould and delays may cause the material to congeal and prevent further pouring. Pouring should begin at the end funnel and the compound be allowed to flow from the air holes for a little while before they are plugged with clay. The material flowing first from each hole will contain entrained air and if allowed to remain will form a soft, spongy, honeycomb pocket in the back of the mould.

Pouring should not begin at the next funnel until the material has reached it and started to rise up it, otherwise an air pocket will be formed between the two funnels unless there chances to be a convenient air hole. Honeycombed material and air pockets will cause the face of the finished mould to sag, thus forming a bump on the cast which, on enrichment, will be virtually impossible to remove. When the mould has been completely flooded the funnels should be topped up to maintain the pressure and act as reservoirs for replacement of shrinking hot melts (see Fig. 71). When the compound has set the funnels may be removed, the compound cut flush with the case and the case released by removing the wads. The case may now be lifted clear of the model, turned over and placed on the bench. Because of the enrichment, the model is usually more undercut than the case and will retain the moulding compound. If, however, it comes away with the case it must be removed.

The case must now be treated – if this has not already been done – to house the compound. Nipples formed in the air holes will not sit back in these holes and will have to be removed. Being so small, they may be cut back behind the line of the compound without causing it to sag. Nipples from the larger pour holes, however, must be cut flush with the back of the compound. With hot melts this may be done effectively by ruling off the cut with a heated joint rule held in a pair of pincers. The pour holes in the case are also made good as both hollows on the back of the compound and holes in the case will cause the face of the mould to sag. The strike-offs and sides of the case need to be seasoned to suit the material to be cast from the mould. Where necessary, the pliable material is also treated for casting. (See Fig. 70 (b).)

See also entries under specific moulding compounds and casting materials.

RUN AS A RUN CAST

To case mould with a run-cast case the method is as described above, but the case is produced as a run cast. The metal profile is set out to the contours following the section and is made to contain correctly splayed strike-offs that will sit on the model's moulding grounds (Fig. 72 (a)).

A limitation of this method is that, to combat the force the case will be subjected to during pouring and subsequent use, it must be adequately reinforced with cross-brackets and long bearers. 25×50 mm timber completely imprisoned in a run cast will give problems of splitting and twisting. It is, therefore, advisable to use metal, e.g. scaffold tubes for main bearers and angle iron for cross-brackets. Producing a case in such a way cuts out an operation only when the section is sufficiently flat. A model with a deep contour, such as an enriched cornice or entablature, will have to have its case run over a core. See RUN CASTING. For use see BEAMS.

Advantages of the method are that, when the case can be run on the bench as a reinforced moulding, only one operation is needed; also the profile may be re-used to run a curved case directly (Fig. 72 (b)). Using the run-core-method above, a curved core will have to be run off with the profile and a curved case cast from it. (See Fig. 70 (c), (d).)

Curved run case

Where no straight moulding is required the model may be run curved, dressed, a core run over it, the case cast and the mould poured, every stage being as for run case (run-core method). If, however, the model is a cornice or entablature that will have to be stood up and run in its projection (see Fig. 31), the head of pliable moulding material needed may require special funnels or may even prove impossible to pour. In this event the model will have to be run down straight, moulded, and the pliable moulding material placed in a curved case. In all instances the arrises in the case should be designed so that they will act as lips and ledges holding the compound on to the case. An extra factor influencing the choice between the two methods discussed above is that a cast case is portable and easily stored.

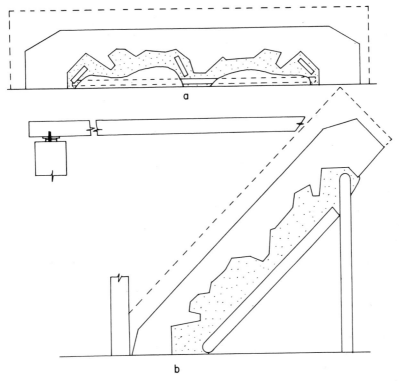

Fig. 72. Mould, reverse (fibrous): (a) a run-cast case for a straight mould, (b) a spun run case

Flood moulds

A mould made by pouring sufficient moulding material over an open model, ringed by a retaining wall, to flood it completely, covering the highest parts to form a flat back to the material. The obvious limitations are the size of the model and the height of its highest parts; these govern the amount of material needed. Fewer operations are involved in producing this type of mould but its chief advantage is that it can be curved by laying it over a DRUM or by bending it to lines.

The walls are erected some 40 mm or so away from the model in order that the moulding ground will form the strike-offs on the mould. If a back is to be cast over the mould, the walls should be splayed inwards to allow it to draw. The choice of the most convenient material for the walls will depend on the shape and height of the model. Suitably seasoned, planed timber can be used for straight walls, clay being used to seal the joints. However, clay beaten and cut into strips and/or lengths of flexible busk are used for curves. The necessity for a back on the mould will depend on the material used and the shape of the mould. Long, narrow moulds, straight or curved, will need a back to hold them to shape.

For the seasoning of the model and mould see under the specific moulding material.

Wax. These are relatively small moulds and, although shrinkage takes place on cooling – especially in the thicker areas – as the material is virtually rigid it does not require a back.

Gelatine. The back of the material will be abso-

163

lutely flat and, therefore, if its shape is such that it will not distort or if it is to be bent to lines, it will not require a back but is merely laid on a seasoned surface.

PVC. This material shrinks substantially on cooling, sinking in the thicker parts and leaving a ridge round the wall. The texture of the back may be rough, due to bubbles having erupted through it. Therefore, with all but the smallest of moulds, a back will be necessary. If the mould is to be laid over a drum or bent to a line, a further batch of PVC may be poured over the first when it has set fully. Being poured on PVC, its back will be perfectly smooth and the shrinkage hollows on the first pour will be almost eliminated. To cast a plaster back the lip round the walls is first cut off with a knife held at about 45° across the angle. Notches may be cut at intervals along the edge to locate it in the back, as the PVC is prone to stretching and shrinking. If the plaster forming the face of the back breaks through the bubbles on the back of the PVC the resulting incrustations should be cleaned from the cast before it is seasoned in preparation for receiving the PVC.

Cold pour. The back of this material will be smooth and flat and will require a back under the same conditions as gelatine.

Insertion moulds

An insertion mould is one consisting of a rigid material containing strips or areas of pliable material. The rigid material both produces parts of the casts taken from it and acts as a case to support the pliable moulding material. The pliable sections may be formed by pouring them as flood or skin moulds, clay-case moulds or run-case moulds. The advantage of an insertion mould over a full case mould is that the portion reproduced by the rigid part of the mould has good lines. This is particularly advantageous with run mouldings where deviations and bumps will be most noticeable. Lengths of mouldings, containing lines of enrichment, produced from such moulds appear crisp and clean in comparison with those from full case moulds.

1 To form an insertion by flooding or skinning the enrichment a model is laid down in the usual way, walls or fences erected around the enrichment, and the pliable moulding compound poured as for a flood or skin mould. The walls should be splayed inwards in order that no undercut is presented by the compound to the back of the case. A plaster back is now cast over the whole model in the usual way according to the materials being used. The back will support the pliable sections and act as a cast reverse mould for the remainder of the model.

2 To form this type of mould with the pliable sections in the form of a clay-case mould, the areas are prepared and clayed as for clay-case moulding. A back is cast over the whole in the rigid material which will act as the case for the compound and as a cast reverse mould for the rest of the area. When set, the cast is removed and the procedure for pouring is as for clay-case moulds.

If a cold pour is being used, which facilitates the sealing with plaster of the join between the case and the model, or, with hot melts, if the rigid mould occupies a relatively large area compared with the cased section, the mould will probably not fit back snugly over the model because of the amount it has expanded on setting. Although a special non-expansion plaster may be used, if it is feared that the mould cannot be forced down tightly to the model a case may be formed separately over the core. The clay-case mould is formed and poured over the enrichment in the usual way, care being taken to protect the rest of the model with paper. The case is seasoned and oiled so that with careful removal the pliable material is left in place on the model. After removal of nipples and pour holes the rigid mould may be cast over the whole.

3   Several variations exist for forming the insertion by the run-case method. (a) A *plaster core* may be run over the model's enriched section only and the case cast over the whole, thus forming a cast reverse mould for the rest of the section. On setting, the case and core are removed and the procedure for pouring continues as for cast run-case moulds. (If a cold pour is being used or the mould will not fit back snugly over the model, see second paragraph of (2) above.) The added advantages of the cast insertion mould are that it may be cast in any hard casting material, is portable and may be stored.

(b) By forming a *run-cast case* that will fit over the enrichment that is laid down separately, the case may be let into a formed reverse mould during its manufacture, or the profile that ran the case may be superimposed on a profile to run a plaster reverse mould, thus producing a bed within it to receive the pliable compound. Alternatively, this run-cast case may be placed over the enrichment on a full model in order that the compound be left behind when the case is removed, allowing the reverse mould to be cast over the whole as in (a) above.

(c) A method developed to cope particularly with *enriched lengths* of mouldings. It allows a strip of pliable compound to be fitted into a bed formed in a run plaster reverse mould. The original profile for running the reverse mould, with the aid of a few metal strips, can be utilised to run a case and core for the production of the pliable strip. The profile for running the mould is cut to the section for a reverse mould but made to include the beds for the enrichment as for a model (Fig. 73). When required, the reverse mould may be run off in the usual way with the plate '3' forming the bed for the strip. For uses see ARCH; BEAM.

Forming the strip. First the two strips of metal '2' are added and the case '*A*' run to the required length. The bed is shellacked and greased, plate '3' removed and the core '*B*' run up over '*A*'. The exposed surfaces of '*A*' and '*B*' are shellacked and greased, casting rules erected, and the model '*C*'

cast on top. The model '*C*' is then removed and the enrichment laid in its bed. The case '*A*' is now used to pour the strip as for run-case moulding. When cool, the strip will fit into the bed in the reverse mould formed by plate '3'.

The added advantage of methods (b) and (c) is that the plate that formed the bed for the compound may be incorporated in any profile to run a reverse mould in any position, curved or straight, flat or over drums. An advantage of run insertion moulds for reproducing lengths of enriched moulding is that the strips may be stretched or shrunk independently in order to balance the enrichment at mitres and pick-ups.

Skin moulds

Hot-melt PVC

These are poured as open moulds as with flood moulding, except that the moulding material forms a skin over the high parts of the model rather than flooding to a flat surface. Such moulds need to be supported by a fibrous plaster back. PVC is a good material for skin moulding as its surface chills and solidifies immediately on contact with the model. Care should be taken to cover each projecting high point quickly so that seams are avoided. For seasoning of the model see POLYVINYL CHLORIDE.

Although undercut detail that is submerged in the low pockets can be moulded, undercut sections projecting above the flood level cannot be skinned for two reasons:

1   The vertically falling PVC will not coat them under their overhang.

2   Any part of the plaster back following the PVC into the undercut cannot be withdrawn from the model.

Immediately on completion of pouring, excess molten material flooding the low parts can be lifted with a small tool or gauging trowel on to the high

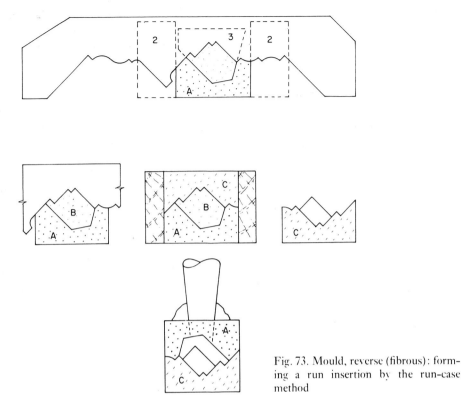

Fig. 73. Mould, reverse (fibrous): forming a run insertion by the run-case method

points, so building up the thickness of the skin as it congeals. Care should be taken not to disturb the solidified face by digging too deeply or for too long after the material has substantially cooled. The fibrous plaster back is cast over the PVC when it has cooled.

Cold pour

A thixotropic grade of cold pour is available. This is applied over the model with a brush. Models should be seasoned as for cold pour. If there is very fine detail, a coat of standard cold pour may first be brushed on to the model and, when tacky, be followed by the thixotropic grade. The material is built up to the required thickness in 3 to 4 mm thick layers. Canvas may be used to reinforce the thixotropic grade and is best laid on immediately and worked in with a brush directly after it has been applied. The fibrous plaster back cast over the mould can be made to bond with the cold pour by leaving a layer of canvas only half worked in. Alternatively the cold pour is left smooth and seasoned so that the plaster back can be released. Plaster backs that have to cope with the undercut presented by moulding in the round are cast in pieces as with clay-case moulds (multi-piece).

MOULD, RUNNING/HORSE

A framework of braces and runners holding a profile rigidly in position so that it may be passed over a setting plaster mix along exactly the same course each run.

Basic construction

The profile is stiffened by fixing it to a stock cut bigger all round to give 3 to 5 mm clearance. The stock is fixed to the centre of a slipper cut at least $1\frac{1}{2}$ times the length of the stock. The profile is fixed to the leading side of the stock, i.e. on the left for running right to left and spinning clockwise. All other components (nib, slipper, struts) serve to hold the stock rigid and prevent chattering.

Benchwork

As in Fig. 74 (a), a small section will require only one slipper and one strut. The best way to stop the nib twisting or vibrating is to add a nib slipper (b). Vibration causing chattering need be only 0.5 mm. A nib slipper will also ride over local undulations in the running surface and will allow extra braces

to be carried. To combat bending in long or deep stocks (c), brace from the back and, if necessary, also from the front. Figure 74 (d) shows a running mould with a special strut to carry the stock braces. In (e), although the bottom of a main slipper is held rigid by the running rule, the forces carried by the struts to the top of the slipper will cause the slipper to bend in its depth, allowing the nib to swing between left and right. This may be combated to some extent by strutting from the extended main braces on to the bottom of the main slipper.

Running moulds of this size will inevitably contain movement and call for extra technique in running. These sections will be cored out with a muffled run. The section may then be run up, using the suction afforded by the core, before any set has taken place, thus allowing the run to be finished before the expansion gets fully under way. FUR-RING may occur initially but is eliminated in the normal course of obtaining the finish. See also SPIN-NING; DOME; RUNNING CIRCULAR CURVES; RULE, ECCENTRIC; RUNNING CHANGING CURVES; CORNICE;

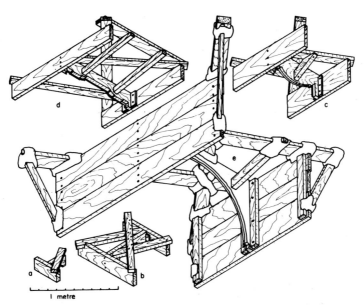

Fig. 74. Mould, running (fibrous): basic running mould construction

MITRE, RUNNING UP/MULTIPLE RUNNING; RUNNING DIMINISHED; NICHE.

MOULD, SULPHUR

One of the original methods of producing reverse moulds from plaster models for pressing COMPOSITION. They are made by melting sulphur sticks over a low heat. Continual stirring is essential to prevent the sulphur from burning.

Plaster models are prepared as for WAX MOULDING. The sulphur must stand and cool to approximately blood heat before being poured. At this time a skin will form over the sulphur; gently break this at one point so that the liquid will pour freely over the model. Allow the mould to cool for a few moments then gently place iron filings into the back. This will strengthen the mould against pressure from the press. Remove the mould before it contracts by cooling and cast a flat hard plaster back over it, again to provide resistance against the press pressure.

When re-melting the mould, remove the iron filings from the liquid by ladling or straining. Again continual stirring is essential, otherwise the iron filings may form into one lump.

Do not allow the melting sulphur to boil, this will cause brittleness. Should the liquid catch fire, cover the pot immediately with a close-fitting lid.

MOULDING GROUND

See GROUND, MOULDING.

MOULDING, TO ENLARGE OR REDUCE

The projection and the depth are divided into

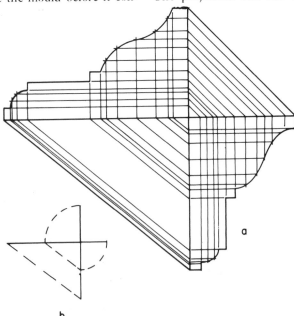

Fig. 75. Moulding, to enlarge or reduce (geometry)

ordinates (Fig. 75(a)). All the members and points round curves are projected on to the projection and the depth as shown. The new projection and depth are marked on the corresponding right angle and each respectively connected to the projection and depth of the moulding. Each scale is altered to the new size by the principle of similar triangles. Each ordinate is drawn parallel to the hypotenuse. The new moulding is then plotted from these.

To decrease the projection proportionately to the decrease in depth, as in Fig. 75 (b), the projection is marked down the depth, then drawn on to the new depth, parallel to the hypotenuse, to form similar triangles.

MOULDING, SECTIONS

See Fig. 76.

Ovolo

(A) Quarter-round comprising a quarter of a circle.
(B) Segmental spun from a point swung as an equilateral arc from both ends.
(C) A parabolic curve plotted by dividing the height into the same number of equal parts as the width.
The ovolo may also be elliptical and set out as (H).

Cyma recta

(D) Segmental. Two equilateral segments whose centres are each spun from the centre and one end.
(E) Each curve a quarter-circle with its centres on the same horizontal line.
(F) Each curve parabolic set out as (C).

Cavetto

(G) A quarter-circle.
(H) Elliptical set out by dividing the depth into the same number of equal parts as the width.

Staff bead

(I) Three-quarters of a circle.

Scotia

(J) Comprising two-quarter circles, one twice as large as the other.
(K) Elliptical set out by dividing the centre line of the rhombus into the same number of equal parts as the projection of the moulding.

Large torus, small astragal

(L) Half circle. The torus usually appears in the base of columns, pilasters, etc., and the astragal in their necking.

Bird's beak

(M) Two curves intersecting to form a beak-like arris.

Cyma reversa/Ogee

(N) May also be set out as (D) and (F).

Corona

(O) The outer flat vertical member of a cornice.

Drip

(O) A hanging member designed to cause water running down the face of the moulding to drip off, thus protecting the wall below.

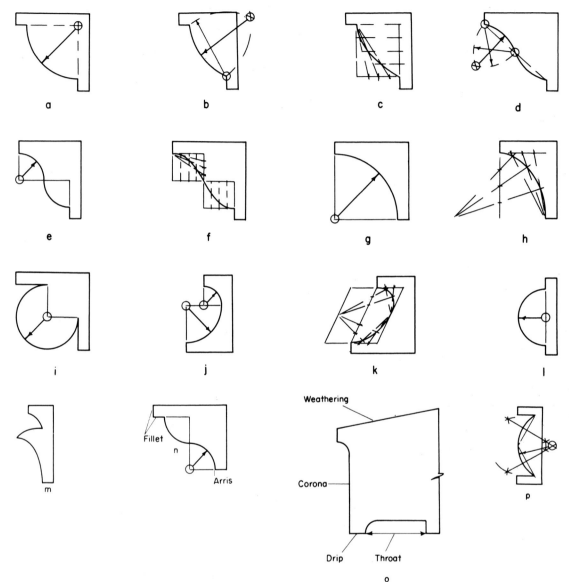

Fig. 76 Moulding, sections, (Geometry)

Throat

(O) The groove in the soffit of a moulding to prevent water running down the face of the moulding.

Weathering

(O) The sloping top to a moulding, designed to cause the water to run off rather than gather and soak into the wall and moulding.

Flush bead

(P) The segment finishing flush with the surface is plotted as shown.

MUFFLE

A muffle is a false profile fixed over the metal profile on a running mould. In both bench and solid running, it is used to run off a plaster core some 5 mm smaller all over than the running mould's profile, leaving just 5 mm of plaster to be applied during the running off to build up the section. These cores are well keyed and become an integral part of the finished run. In bench work only, a muffle may be used to run a core that is shellacked and greased so that a run cast may be made over it or a run case cast off it.

The false profile may take many forms. Some of the most common are:

1   A profile cut from sheet material, such as metal, plywood or hardboard.

2   A plaster muffle may easily be made to follow complicated sections and dispenses with the necessity for involved cutting. To attach such a muffle to the stock, nails are driven halfway into the stock behind the profile. A plaster-soaked strip of canvas may be poked round the nails to reinforce the muffle. As the plaster sets it is modelled with a small tool to follow the shape of the profile. The muffle should be shellacked and greased before running.

3   On large plain sections, both curved and straight, layers of 6 mm lath may be tacked behind the metal profile to extend the stock.

See also CORE and the particular item to be muffled.

MULLION

Vertical member used to divide windows into separate lights.

MUTULE

Projecting inclined block in the Doric cornice, usually placed over the TRIGLYPH.

NEAT

Plaster or cement gauge, but with nothing added but water.

NECKING

The area between the astragal of the shaft and the start of the capital proper in the Roman Doric order. Can also apply to other capitals.

NEWTONITE LATH

See BACKGROUND.

NIB

See MOULD, RUNNING.

NIB RULE

See ANGLE, EXTERNAL RUN, internal plastering; CORNICE, SOLID, EXTERNAL; RULE, eccentric.

NICHE, FIBROUS

A niche is a shallow recess in a wall, usually to take a statue or a vase.

A reverse mould for forming a niche may be run, formed, or cast from a run or turned model. The choice of method to produce the mould will be restricted to the processes producing the required combination of head and shaft sections. All the methods of forming niches *in situ* can be employed, in conjunction with a running or turning box, to produce a model from which a mould can be cast. One of these may be chosen, either as a quick and accurate way to produce a certain head and shaft combination or in order that a portable and storable one-piece mould may be cast from it.

Usually only the shaft and head of a niche are cast – the base is formed on fixing, in either plaster, timber, glass, or some other material. The mould, therefore, is made with a shaft longer than the required cast, to accommodate a casting rule that will form the end of the niche. As subsequently described, the wall containing the niche can be cast as one unit with the niche, while other features –

such as mouldings round the head, enriched heads, and provision for hidden lighting – may be cast as integral parts of the niche.

Run reverse mould (semi-circular head with semi-circular, semi-elliptical or other shaft)

Figure 77(a) illustrates how a sufficiently small niche may have its shaft run up in one. The stock is then cut through the centre of the section, a shoe fixed to the newly-formed nib, and the head spun from the pivot block. The head is then cut off across the centre pin to form an exact half-dome, the shaft placed to it, wadded up and made good.

A larger niche (b) may have its shaft run up in two halves as a RUN CAST formed over a shellacked and greased CORE, the running mould altered and re-horsed to spin the head. The head may be cut accurately to the half-dome and left in position or, if run up over a shellacked and greased core, may be removed for assembly. The pieces are then set up, wadded together from the inside and made good.

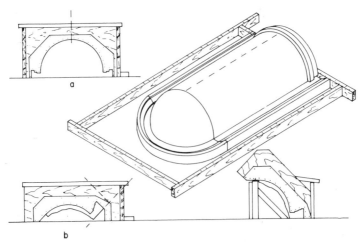

Fig. 77. Niche mould (fibrous): run-cast shaft with spun head

The centre-rule method (Fig. 78) produces the same combination of shapes as do the methods illustrated in Figs. 77(a) and (b), but allows the mould to be run up in one. A rod on the nib of the running mould fits into the metal channel guide on the centre rule. The mould is run from right to left. As the rod meets the stop at the end of the channel the head is spun, the slipper leaving the first rule and engaging with the other. The running rule on either side is fixed to the inside of the slipper, a suitable hole

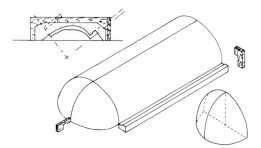

Fig. 79. Niche mould (fibrous): spun head to fit a run-cast semi-elliptical shaft

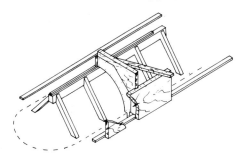

Fig. 78. Niche mould (fibrous): run up in one – centre-rule method

being notched from the stock to accommodate this, so that the slipper may leave one rule and engage with the other when turning round the head.

A shaft of any section may be produced by the three methods described above, its section being determined by the metal profile, but the head must, of course, be semi-circular. Any form of arris, moulding and rebate for jointing with surrounding plaster may be incorporated by including the necessary shape in the section.

Run reverse mould (semi-elliptical head with semi-elliptical shaft)

The shaft may be run off with an elliptical profile, as illustrated in Fig. 79, using either of the methods outlined above, and the profile adapted to spin a semi-circular head. The head is cut to a true half-dome and then cut in half again. The quarters so

formed are then rearranged to produce the elliptical head as shown. A head so formed will naturally contain no strike-offs and these must subsequently be produced by running them with a (profile) gauge off the head blocked up to the correct height to correspond with the shaft.

Mould cast from a run model (semi-circular shaft with semi-circular, semi-elliptical or other head)

A niche with a semi-circular shaft may, of course, have its shaft formed by method 1 or 2 and a second profile cut to produce its head by method 4, but a model may be turned in one as would a column mould. The shape produced, however, will be a model without the necessary grounds to produce

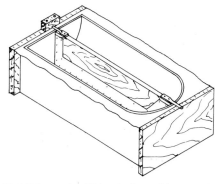

Fig. 80. Niche model (fibrous): formed in a turning box

strike-offs on a mould cast from it. The moulding grounds may be provided by fixing casting rules down both sides of the shaft and running up with a gauge round the head. Although the shaft is run up with a wooden profile and finished as a column mould would be, an effort should be made to form the head with a metal profile. (See Fig. 80.)

Mould cast from a run model (semi-elliptical head with semi-elliptical shaft)

A model for such a niche may be run up in a box as the method in Fig. 81 and a mould cast from it.

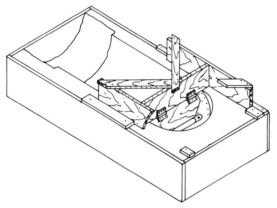

Fig. 81. Niche model (fibrous): run up in a box with a hinge mould

Grounds to form the strike-offs on the mould may be run down both sides of the shaft on the model or formed by casting rules, but the ground round the head can be run up afterwards with a gauge cut to the required section.

Run and formed mould

The heads of really large niches may be cast and made up as would DOMES, either as one unit or in sections. The shaft may be cast in sections as curved plainface from a DRUM and made up on fixing as with barrel work.

Enriched niche

Niches containing enriched heads (e.g. scallops) may be made up in several ways. A base ground on which to model the head will have to be accurately formed by one of the methods that will produce the required head as a model. The head may be formed on a model containing a shaft off which a clay-case insertion mould may be taken – pliable material for the head and rigid for the shaft. This type of mould will allow the niche to be cast in one – a considerable saving of time when many are to be produced. The other method is to clay-case the head, and cast it separately. It can then be joined to the shaft either when fixing, or by wadding it to the shaft before they leave the shop, or by placing it against the end of the shaft mould and stitching it on as the shaft is cast. Alternatively the niche may be cast in one by placing the case mould against the end of the shaft mould.

Surrounds

A niche may be cast containing a portion of the surrounding wall. A rectangular cast can be made to pick up more easily with other fibrous plaster plainface, coincide with the fixings for expanded metal lath, or provide the entire wall surface where the niche is to be contained in a false wall across a corner or an existing recess. In the case of a plain arris the niche mould is formed on the bench with no strike-offs, in order that the bench will produce the wall face. When a moulding runs round the niche its reverse section is run up with the niche, and the bench surface raised afterwards by ruling in off the moulding ground on the moulding (see Fig. 77).

Mouldings round semi-elliptical heads

To a model. A model for a niche with a semi-elliptical head formed in a running or turning box by method 5 or 6 may have an archivolt-type moulding

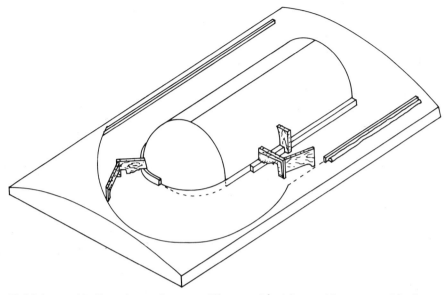

Fig. 82. Niche mould (fibrous): running a moulding round a niche mould on a curved background

run round the head quite easily, working the end of the slipper off an eccentric rule and using the rim of the niche as the buried rule on which the nib of the running mould will hook.

To a reverse mould. A reverse section for such a moulding can be run round the reverse mould for a semi-elliptical head on buried and eccentric rules as shown in Fig. 82.

Independently. Elliptical mouldings may be cast separately and fixed round the niche afterwards.

Hidden lighting

In order to conceal lighting, the wall may overlap the niche. This may be achieved by casting a plainface containing a hole to the correct shape, but a size smaller than the niche. The cast niche is then placed in position on the back of the plainface and wadded securely; alternatively the niche may be fixed first on site and the plainface fixed over it after the electrical wiring has been completed. Concealing the lighting may also be achieved by the addition of an overlapping moulding fixed round the opening of a plain arris niche.

Curved backgrounds

When a niche is to be fixed in a curved wall, the mould will need to be formed on a DRUM that has been ruled in from ribs. Probably the easiest way is to form a mould for the niche, place it on the drum (blocking it symmetrically about the drum) and run up the strike-off with a gauge. Any moulding required round the niche can be run on the reverse mould using a three-legged peg mould run on an eccentric rule and buried rule (Fig. 82). When the niche is to be cast containing a portion of curved wall the head of the mould for the niche may be trimmed slightly concave to fit the drum, the mould fixed on the drum and made good.

Casting

The niches are cast in the same way as curved PLAINFACE and DOMES with bruised laths flat over

175

the curves or knuckle-jointed laths on edge where room exists.

Fixing

The casts are nailed or screwed to timber or wired and wadded to metal, as with all large fibrous plaster casts. See PLAINFACE, fixing; VAULTING.

Jointing

Because the soundest possible joint is one that is reinforced with jute scrim, provision should be made round the perimeter of the unit to provide this with the adjacent plasterwork, e.g. rebates to pick up with other fibrous plaster casts, or floating and setting rebates to pick up with solid work.

NICHE, SOLID

When forming in solid, one should consider the size as well as the shape before beginning work. Some niches are too large to run and they are formed by using pressed screeds, and floated and finished in this way. Others are run up with a horizontally worked running mould. This stops at the springing line and the construction of the running mould allows the head to be spun by drawing the hinged part of the mould towards the front. Smaller niches can be turned by a vertical running mould fixed centrally to the niche width on the wall face. Shallow niches may be run up on a central rule, again stopping at the springing line and the head being spun from this position.

Solid, pressed and spun screeds

For very large niches the procedure is similar to floating to curved surfaces, the body of the niche having either dots and pressed screeds, or run

screeds at the top and bottom. From these the body could be floated, and external angles formed where the niche joins the wall surface. The head could be formed in the same way; however, if possible it would be preferable to run the external angle in much the same way as one would run an arch. The profile would consist of an external angle, straight on the wall side and following the niche-head curve on the niche side, about 50 mm either side of the arris. This mould would be mounted on a gigstick to the head radius, a bearer or stretcher being wedged across the opening and a pivot block fixed to the centre. The profile could continue on the head side to produce a screed for floating and of course a setting member. When this has been run and the body floated, the head can be ruled in from these with a purpose-made rule, cut to the curve.

For elliptical niches worked in this way a peg mould and rib would be used instead of a gigstick mould. Where the niche is surrounded by a moulding, the vertical section could be run in the normal way and the head spun as in running moulded

Horizontal running mould

This method can only take place when the niche body and head are both the same shape. A double-horsed running mould is constructed, the slippers being one on either side of the stock. The profile is cut to the shape of the niche in plan and the mould is horsed so as to allow the stock and profile to pivot upwards only. Two or three hinges may be used to do this and additional support should be given at the back of the stock to prevent it dropping.

The best way to construct this mould is to cut two stocks or half-stocks. The first, to back up the metal profile, should be cut accurately along the diameter line of the niche. The second or rear stock should be at least 150 mm wider than the niche, and its own width should be sufficient to take the two slippers.

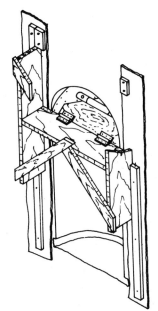

Fig. 83. Niche (solid): running a niche with a horizontal hinge mould

Once the metal profile has been fixed to the first stock the two stocks can be hinged together. Mark a centre line on both stocks and fix so that these lines meet. Fix with the hinges so that they will open in an upwards direction when the mould is placed in position. Turn over and fix a strong support to the running stock that will take the weight of the profile stock as the mould runs upwards, then fix the two slippers and cross-braces. Finally make a wire lifting handle and fix centrally to the profile stock.

The positioning of the running mould and the provision of screeds will be as for SCREEDS, running in lime, putty and plaster.

Alternatively it is possible to run on two rebates, one fixed to each slipper. Once the rules have been fixed, the running mould is held in position at the springing line and two wooden stops are fixed to the wall at the point where the front of the slippers rest. These will effectively stop the running mould from going past the springing line when the niche is being run. A T-strut should also be constructed, the height of which must be sufficient so that it can be placed underneath the rear of the slippers to take the weight.

Two operatives will be required to run this mould – feeding and running are both usually two-handed jobs. As the mould passes up the niche, it will gather a certain amount of material. When it hits the two stops, one man should place the T-strut in position while the other steadies the mould. Then he should thoroughly clean the profile. While one man continues to hold the mould steady the other should grip the wire handle and pull the profile towards him. Provided sufficient material has been applied to the head a few turns backwards and forwards should get the required shape, and one finish run should complete the exercise.

The head arris is formed by working the mould backwards at the edge. All surplus material should be cleaned off with a joint rule. The floating or backing to the finish, if required, is formed by the same method using a muffled mould. The running rules should proceed well beyond the base of the niche, to ensure that it can all be run. The side or vertical arrises will be fixed during the running of the niche body. A seam will probably occur at the joint between the body and the head but this can be cleaned off with flexible busk when the job is complete. (Fig. 83.)

Vertical running mould

The entire niche is formed in one operation but the method has one main disadvantage in that it does not produce completely run angles with the wall face. Its main advantage is that an impost, or similar moulding continuing around the curved surface of the niche, can be run up at the same time.

The wall face should be floated first, and a metal profile cut to the complete vertical section of the niche with or without impost mouldings and mounted on a stock. On the wall surface, central to

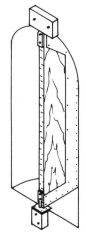

Fig. 84. Niche (solid): turning a niche with a vertical profile

the niche elevation, a line is plumbed down. At these points a bolt catch, slot or the like is firmly fixed, both top and bottom, the two bolts are fixed, to the back of the mould stock, one to slot into the head catch, the other into the base catch when the running mould is held in the correct position, and both bolts are rammed home. The mould should pivot on these in the same way as a door.

It is not essential to have bolts and sockets for this operation. One could quite easily make up a set out of metal rod for the bolts and soft metal for the clasps. However, it is essential that both bolt sections should be easily movable when the mould is removed, yet very firmly held when the running is proceeding. If wooden dowels are used they will expand when they get wet, and may splinter or even split under pressure.

Once the two pivots and bolts have been fixed the mould should be muffled for floating. The plaster is applied evenly to the entire area and the running mould placed in position. A wire handle should be fixed at least to the profile side of the mould and then it can be pivoted on its centres using this.

Continue till the backing is perfect, then remove the mould and clean; also key the backing. After the

muffle has been removed, apply the finishing coat in the same way. In this operation, the running mould may have to be removed from its pivot after each run so that it may be properly cleaned. Run out well past the external arrises in every case, and when the niche interior is finished, these may be trimmed so as to act as setting members for the wall surface. (Fig. 84.)

Centre-rule or screed method

Arris with a moulded surround may be run in one operation. Or in the case of a plain surround all external angles are run up, thereby forming good setting members for the wall finish. In general, however, this method is only used when forming shallow niches on plan, while the head remains semi-circular in elevation.

The wall face must be floated first. A metal profile

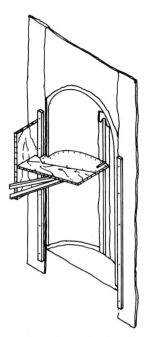

Fig. 85. Niche (solid): running by the centre-rule method

cut to a half horizontal niche section, with or without the surrounding moulding, as appropriate. The profile should be horsed in such a way that the slipper will bear on the wall face. A bedded flat rule or floated vertical screed is then applied to the niche background, right in the centre. On this the running mould nib will bear. At the top and central to the niche head a pivot should be fixed, complete with projecting head. (Fig. 85.)

Lime-putty and plaster screeds are applied to the wall face at the slipper points, including the head, and if a floated screed is being used in the centre, to this also. Rules are then fixed plumb, so that the slippers will bear on these, and the two sides are run up on a muffle run core. The rules are then removed and the head spun, using the pivot as the centre, and a fishtail fixed at the nib position of the running mould.

It is possible to run a niche of this type in one complete operation. To do this the side running rules must be on the inside of the slipper and cut to the exact springing-line height, also the pin will be on the running mould, the fishtail on the centre rule. One can then run up, usually the left-hand side first.

Make sure when the mould reaches the springing line that the fishtail embraces the pivot. Then spin round the niche head carefully and check that the slipper contacts the right-hand slipper rule correctly and continue running down on the right-hand side. This is quite a complicated operation when aiming at perfect results. Therefore one should run a clean mould around the niche several times before actually running the niche. This is to check that the rules are right and that the pivot will fit directly into the fishtail without any movement. Accuracy in setting out is essential.

Where one has used a bedded or buried rule for the nib bearing this should be removed and the gap made good. Where a screed has been used, the running joint must be cleaned off and made good.

Summary

It will be seen from the foregoing that the formation of niches in solid plaster can be quite a complex question. Which method to use, for instance? In most cases the decision will be left to the craftsman on the spot, in view of his likes and dislikes plus, of course, his practical experience and ability. A rough guide could be pressed screeds for the largest, horizontally run for the next largest; and vertically spun for the smaller niches or those containing a horizontal moulding. This would leave the centre-rule method for the shallow and full-moulded surround.

However, this must only be a guide and the choice will be limited to a combination of shape. Many plasterers will run even the largest niches using a vertical spinning mould. Also one must be aware of the numerous permutations of niche shapes: semi-circular head with segmental body, and elliptical body with semi-circular head, a deep body with a shallow head, and so on. For all the methods given here it should be possible to form almost any curved niche, as in many cases what one will use is not a particular method but a combination of any two.

NOGGIN PIECE

Timber cross-member between the main members of a timber-framed construction. May be wedged between the flanges of a BSB so that a fixing may be secured.

NOSING

See STAIRCASE WORK IN GRANITE AND CEMENT.

OGEE

See MOULDING, SECTIONS.

ONE-GAUGE

See CASTING IN FIBROUS PLASTER, one-gauge method.

OVAL, PLASTERER'S

See ELLIPSE.

OVERDOOR

An ornamental feature over a door opening. It may be a pediment or take the form of a lintel as an entablature returned at both ends.

OVOLO

See MOULDING, SECTIONS.

PAINT, REMOVAL OF OLD

See REPAIR AND RESTORATION.

PANEL MOULD

A length of moulding cut and mitred to delineate a panel.

PANEL, PLASTER (SOLID), PLAIN OR MOULDED

Plaster panelling, often referred to as *fielded work*, may be plain or moulded. Plain panelling can be formed in varying shapes and the panels will be either raised or sunken. To form this panelling the surrounding surfaces should be floated to a true surface first. In the case of sunken panels, the panel area should be cut back from the wall face and the sunken area ruled in with a panel gauge.

Alternatively, once the wall surface has been floated, thickness rules may be fixed inside the panel measurement and the outer surfaces built out to these. Raised panelling will be formed in the same way, with the thickness rules fixed, this time outside the panel measurements. The raised panel surface is then built out to these (Fig. 86).

To finish, remove all rules and set the lower surfaces, grease and replace the rules keeping the inner edge 2 to 3 mm away from the raised section. Set this and push plaster hard into the gap between the rule edge and the raised floating. When the thickness rules are removed, perfect arrises should result. In both the floating and the setting, the thickness rules will act as grounds for ruling off. Curved arrises to plain panelling must be run up, the profile being a plain external angle, and the curve formed by running with a gigstick or peg mould.

The background for moulded panelling must be formed in the manner described for plain panelling. The shape of the panels may again vary, and this time there are three types – raised, sunken and flat. In the case of the latter, they will consist of panel moulding, run or planted on the wall surface. In all cases the running rules must be fixed outside the panel area and the two horizontal lengths run up first. It is normal practice to run after the panel surface has been finished, the running mould nib will bear on this.

At the slipper end it is usual to run on a rebate,

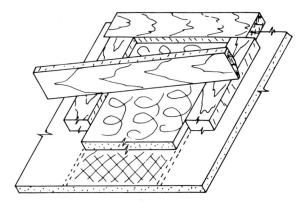

Fig. 86. Panel (solid): to rule in the finishing to a plain raised panel

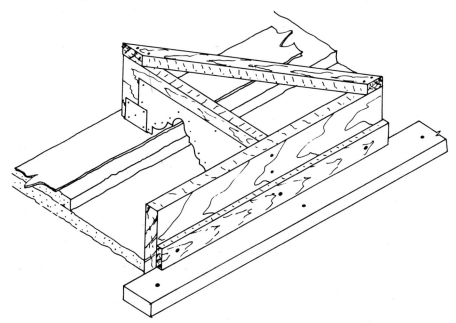

Fig. 87. Panel (solid): running a bolection moulding to a raised panel

thereby eliminating the necessity for putty and plaster screeds. However, a setting member should be formed when running, to act as a guide when setting the outer surfaces (Fig. 87). Large curved lengths must be run *in situ* with either a gigstick or peg mould.

Small straight lengths and small curves may be run down and planted. For all run-down panel work the moulding may be either run on a prepared core, the core representing the raised surface of the background, or run down on a flat background and, when planted, the member that corresponds with

181

the panel reveals should project 2 to 3 mm in front of the floating. This member is then used as a guide to set the panel reveals to.

PARGETING

In modern terms to 'parge' will generally mean to line the inside of chimney flues with mortar. Originally, however, pargeting consisted of solid decorative plasterwork, either in freehand or by stamps, and was carried out by local craftsmen in local materials.

PATERA

Originally a flat, circular, flower-like ornament. Now applies to any small flower, whether circular or not.

PATH

See FLOOR LAYING AND ROOF SCREEDING.

PATTERN STAINING

See FAULTS.

PEA GRAVEL

See AGGREGATE.

PEBBLE DASHING

See FINISH/RENDERING, EXTERNAL.

PEDIMENT

Originally a pediment was a triangular feature formed by horizontal entablature and raking cornices but now also can be ogee or semi-circular in form. It appears at the ends of buildings, as a portico, and over openings. It may be closed or open – with two returns finishing the cornice at the top. The classical proportions, height to span, are $\frac{1}{4} : \frac{2}{9} : \frac{1}{5}$ (mnemonically 'two-eighths, two-ninths, two-tenths').

Fibrous

Benchwork

To run the mouldings to make up the model a metal profile is cut to the full section of the entablature. This will run the two returns. The horizontal entablature can now be run by masking off the top members of the cornice. The raking cornice can now be run by masking off the frieze, cutting away the profile on the top members, and sufficient of the stock to allow a metal profile to be substituted for the raking portion of the section. On open pediments the profile for the top returns will have to be cut and filed afresh. For sections see RAKING.

Plain moulded

These may be produced using piece moulds or clay-case moulds. See MOULD, REVERSE.

182

Casts

May be required in plaster, cement mixes, or reinforced plastics. Whereas for external pediments, moulds could be fixed in position, filled with the appropriate cement mix, and struck when set, it is now more common to cast in glass fibre and fill solid with a suitably reinforced cement mix, either in the shop or in position, according to size. Large pediments will have to be cast in sections and made up on fixing.

Solid

Externally, all large pediments and many of the smaller ones are run up in Portland cement and sand. The actual running of the section will be as in RUNNING, Portland cement and sand. For mitring see MITRE.

Two running moulds will be required, one for the horizontal section and one for the raking section. The first mentioned will be as for running CORNICES in cement and sand with just one possible exception. The return piece for forming the weathering and running on the nib rule will be as small as is practically possible. The raking running mould will be very different. As well as having the raking section of mouldings it must also be horsed so that the slipper is on the outside or above the cornice. This allows as much of this moulding to be run as is possible. If the slipper is put in the usual place, it will hit the horizontal length and make the external mitres much larger. All three lengths must be cored out, then finished in one operation – the mitres are worked in while the running is taking place so that the entire pediment is finished as one.

PEG MOULD

See RUNNING CHANGING CURVES; RUNNING CIRCULAR CURVES.

PENCIL ROUND

See ANGLE, EXTERNAL, FORMED.

PENDANT

Ornament hanging like a stalactite from a ceiling. Often formed by the ceiling itself curving down from four sides like a small vault. Mouldings on the ceiling will form the GROIN lines, covered by a BOSS at intersections.

PENDENTIVE

The remaining triangular spherical surface springing from a support and formed by barrel VAULTING or ARCHES intersecting with a DOME (see Figs 133, 134), but often applied to the comparable formation in cross-vaulting.

PICKLING

See GELATINE.

PICK-UP

1 Of enrichment – continuity at a joint, mitre, etc.
2 Of a mix – to stiffen at the start of the set.

PIECE MOULD

See under MOULD, REVERSE.

PIER

A support, either in the form of a pilaster or of a mass of brickwork between openings.

PILASTER

A rectangular column, usually placed against a wall.

Fibrous

Model

If required, a model is made up by running the shaft, returning any flutes present, running the base and astragal mouldings and planting and mitring them to the shaft. A plain cap will be included with the astragal. An enriched cap will be either clay-case moulded separately or fixed to the model to be clay cased as an integral part of the reverse mould.

Reverse mould, cast

Astragals, bases and plain caps are included in the reverse mould for the shaft. Enriched caps are clay-case moulded and cast separately; alternatively the mould may be placed against the end of the pilaster mould, which has been made to an extra thickness to accommodate this. The piece mould of the shaft and base will comprise two cheeks for the sides and a saddle, reproducing the face. A full model may have the enriched cap clay cased as part of the piece mould.

Reverse mould, run – straight shafts

The section for reproducing the face of the shaft is run, and the cheeks of the mould for reproducing the sides of the pilaster run by masking off a portion of the original profile. If the pilaster is fluted, the appropriate number of flutes (where the cheek will butt against the shaft section) will also be masked off from the original profile. The astragal, base and plain cap are formed by scribing the mouldings run to their reverse sections and placing them against the ends of the shaft. The thickness of each moulding must be such that all form a fluent whole when joined.

Reverse mould, run – diminished shafts

The shaft section is best run with a triple-hinged running mould (see RUNNING, diminished). Diminished cheeks may also be run using the original running mould, masking it off as for straight shafts. The flute/flutes nearest to the centre of the section are used for the flute/flutes on the pilaster's sides. The section is masked off in such a way that the centre line of the stock is the part of the cheek that will be placed against the horizontally straight face of the shaft. The number of flutes required on the pilaster's sides are left on above the centre line and the depth of the cheek is masked off below the centre line to the thickness of the shaft mould. When the mould is put together, to counteract the diminish of the run about the centre line, packing will have to be placed under the cheek as it narrows towards the top of the shaft. The pilaster will have a straight face and a curved back but this can easily be reversed by placing the new cast back downwards on a flat surface.

FLUTES

Flutes on the model will take the form of grooves which are filled in flush to a square line and the returns carved with a gouge or template. Flutes on a run mould will be raised and are returned to length by carving; the portion that runs on beyond the return is scraped off and finished flush with the fillets or arrises. An alternative in each case is to return one flute only, cast a mould from it and form the

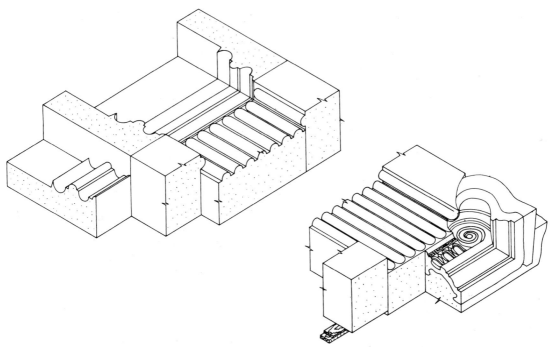

Fig. 88. Pilaster (fibrous): a mould with run diminished shaft with cheeks, scribed run moulding base, and a clay-case mould cap

rest of the returns by casting them against the mould placed in position on the shaft. See also separate entry on FLUTE.

REEDS

These are most likely to be found in the flutes of diminished pilasters. They may be formed in a run reverse mould by superimposing their convex metal profile on the running mould, the two parts of the shaft being run within the same set-up and joined. However, as this makes six parts to the reverse mould instead of three,• it is more satisfactory to make up a model and cast a reverse mould from it.

Casting

Boxed laths down the sides provide edge lath for strength and flat lath for fixing to timber. All longitudinal laths should continue through cap and base to the extreme ends of the cast; otherwise, casting is standard for casts of this shape.

PILASTER, FLUTED (SOLID)

A fluted pilaster may be formed as in columns, using a fluted collar top and bottom. However, as most pilasters have little projection from the wall face, they can very well be run up in position.

For this operation a hinged mould is generally used. Before making the running mould, check whether or not a flute is required on the cheek of the pilaster. Some have just one flute on either side while others do not.

A well-keyed backing should be applied to the background, well behind the finish lines of the pilaster. A temporary batten or line should be plumbed down to the centre line of the pilaster face, and the setting out of the diminishing rules done from this. The procedure should be as in all solid running, setting out lines, strong gauged screeds and re-setting out, this time on the screeds. Twin rules should be fixed either side, one outside each horse or slipper and one inside, the latter to prevent the slippers from wobbling.

The centre batten is removed and the core should be run out to a muffle.

When running using the method suggested here it is advisable to keep the rules and the slipper edges well greased or oiled. The running must be from the bottom upwards, starting off at the widest part of the pilaster with the profile at right angles to the two slippers. Possibly for the lower third of the pilaster the run will be perfectly straight. After this the diminish will gradually begin to appear.

One should attempt to run off in smooth slow runs, not fast erratic ones. When passing the mould up to another plasterer make certain that the take-over is smooth and the pressure on the running mould constant.

The material used in this operation will probably be Keene's cement. Therefore care must be taken to see that no lime putty and plaster screed gets under the Keene's run. Should the pilasters have no backing at all, but merely be run on a flat wall, there is no reason why they should not be run down on a purpose-made bench and planted, using the same method as running *in situ*.

All flute return ends are put in on completion of the running. Form one perfectly, and take a mould from this to assist with the others. This may be done to both ends of a flute on the run pilaster or from a small piece of both sizes run down on the board. See also COLUMN.

PILASTER, PLAIN DIMINISHED (SOLID)

These may be formed in one of several ways. In the first example, plain plaster collars are fixed top and bottom as for columns. An entasis rule is then used to obtain the true shape, muffled for the floating and free for the setting.

The second method is basically similar except that, instead of plaster collars, solid plaster DOTS are used. These are squared around the pilaster, first at the top, then plumbed down as in the collar method. The third method is to include in the casting of the capital and base a minimum of 50 mm of the shaft. These casts are then fixed in position and the backing ruled off with a muffled entasis rule and the finish with a free entasis rule.

With all of these methods, one alternative can apply. In this case, instead of ruling off using the entasis rule, two rules are cut to the same shape. These are then fixed to the sides or cheeks of the pilaster and the face ruled off to them. They are then removed and fixed to the face so that the cheeks can be formed.

When using an entasis rule, there will be a variation of the rule used to form circular diminished columns. This time no slipper will be needed at the ends of the entasis rule.

Finally there is the solid run method. However, this is not often used for plain pilasters, as these are relatively easy to form using any of the systems mentioned here. Most solid fluted pilasters are probably run up and this method will be covered under that heading.

In most cases the background for the pilasters will be formed in either brick, concrete, or the like. If false pilasters must be formed in solid, a core will be needed. This will depend upon the projection, deep pilasters probably requiring battens and EML, while for the slim variety, the core can consist of several coats of well-keyed 'dubbing'.

On a wall containing several pilasters, care should

be taken to line them in so that all projections are parallel. This is carried out in a similar fashion to floating walls containing attached piers.

PIN

See SPINNING.

PINCH ROD

Two rods or rules used together for the accurate measurement of openings and the like when only internal measurement is possible. They overlap and are expanded to fit the measurement.

PITTING

See FAULTS.

PIVOT BLOCK

See SPINNING.

PLAINFACE

A flat, fibrous plaster cast that, when fixed, will form a plain surface such as a wall or ceiling.

Casting

For method of casting see CASTING IN FIBROUS PLASTER, general method, with the following amendments:

Dimensions

In general, when the cast is fixed with others, its size is governed by the fixings: because the cast is fixed across the fixings, its length will be a multiple of their spacing and its width a denominator of the total width. Casts usually vary between 1.830 × 1.200 m and 2.400 × 1.200 m. When fixing to timber grounds, the laths in the cast will be placed flat to facilitate nailing or screwing, but when fixing to metal they will need to be on edge so that they are in the strongest position for hanging on wires. Casts will, therefore, need to be 22 and 35 mm thick respectively. The shape of the plainface is set out on the bench some 5 mm smaller all round than the space it is to fill, in order to provide some 10 mm tolerance between the casts on fixing. Timber casting rules are fixed to the bench along the lines of setting out, with as few nails as possible. (It is only necessary to take up two adjacent rules out of four in order to remove the cast.) Although it is advisable to GREASE the bench under the rules, the greasing of the main area should be the last operation before gauging; otherwise, foreign material – sawdust, strands of canvas, dust, etc. – will stick to the grease and foul the face of the cast. In general, it is not wise to leave the heads of nails protruding to facilitate withdrawal, as, apart from their getting in the way, tangling with canvas and hindering striking off, protruding nails have torn many a finger to the bone.

Provision for jointing

A satisfactory joint cannot be made by butting the square edges of two casts together. The best method of providing a joint that is as crack-free as possible is to form a 25 × 6 mm rebate round the edges so that 80 mm wide jute scrim may be incorporated in the stopping, thus wadding the casts together from the face as well as reinforcing the stopping. This rebate may be formed in the cast by well greasing a 6 mm lath and placing it on the greased bench against the casting rules. When fixing the thicker casts to metal, it is an advantage to have a lap joint

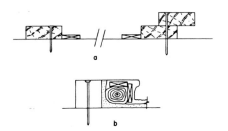

Fig. 89. Plainface (fibrous): (a) casting rules for lapped and rebated joints, (b) boxed laths on a rope

so that one cast may be fixed to the other by wire or a screw, automatically making their faces flush and not flapping one past the other where no fixings happen to be situated conveniently behind them. This lap joint is formed by fixing the casting rules as shown (Fig. 89) (a). (This form of jointing is also invaluable when fixing curved casts to form barrel ceilings, domes, etc.)

Firstings

Firstings is generally applied as described in CASTING IN FIBROUS PLASTER, general method, except that some specifications require that the firstings be ruled off at 6 mm. This may be achieved by striking down with a gauge from the casting rules or by ruling off the laths forming the rebates.

Reinforcement

When using flat laths for fixing to timber grounds, it is possible to place all the laths, i.e. those running both ways, and cover them with canvas at the turning-in stage. If cobwebbing is feared, however, the longitudinal laths only may be positioned and the canvas tucked round them by working in from both sides, pulling the canvas in from the middle. The cross laths may then be positioned and covered with plaster-soaked strips of canvas. It is sometimes preferred to use ropes under flat laths. This means that, with the 22 mm cast, the thickness does not allow continuous laths to run in both directions. The cross laths are, therefore, broken between the longitudinal ones, either at right angles or in herring-bone fashion.

When using laths on edge for fixing to metal fixings the extra depth permits a boxed lath to be positioned round the perimeter of the cast. This is one lath on edge with another flat over it. Where no lap-rule arrangement is being used this is formed by placing a rope in the angle with a lath on edge against it and one flat on top of it. (See Fig. 89 (b)). All perimeter laths are covered by turning in.

Alternative methods of lathing on edge are:
(a) The laths running with the fixings (those across the width of the cast) will not be used for fixing and so may be placed flat and covered with canvas at the turning-in stage. The laths running lengthwise are then positioned on edge and covered with plaster-soaked strips of canvas.
(b) A variation on (a). To gain extra strength, 25×6 mm laths may be cut to produce 10×6 mm laths which are used on edge on top of the continuous flat laths but are broken between the edge ones. These are positioned after the long-edge ones are covered with canvas and are then themselves covered with plaster-soaked strips of canvas. These two methods provide continuous laths running in either direction.
(c) 25×6 mm laths on edge both ways. The longitudinal fixing laths are positioned on edge first and are either covered by canvas at the turning-in stage, tucking them in by pulling slack in the canvas, or positioned and covered with plaster-soaked strips of canvas after the second layer of canvas in the cast has been brushed in. The cross laths, broken

between these, may now be placed in position – either at right angles or herring-bone fashion – and covered with plaster-soaked canvas.

Strike-offs

In general, all strike-offs that are to come in contact with fixings and are to be used for fixing are made up and struck off with a straight-edge rule drawn over the casting rules. In some cases, where a cross lath coincides with a fixing, it may be desirable to make a fixing through it but normally it is only considered necessary to make up the longitudinal fixing laths.

To strengthen and brace the cast during storage and transport, it is sometimes considered necessary to wad 25×50 mm lifting-out sticks over the back of the cast. These are cut off immediately prior to fixing. The battens may more easily be cut off, leaving the wad flush with the strike-offs, by lifting them slightly from the back of the cast. This is achieved by placing them on a 6 mm lath before wadding up, facilitating the passage of a saw between them and the cast.

Cleaning up

The face of a plain cast, whether straight or curved, is usually cleaned up immediately after removal from the mould and while it is still wet. The surface film of plaster and grease is removed by scraping with a sharp 300 mm joint rule and sometimes afterwards with a piece of busk. It is then wiped over with a piece of canvas to remove all traces of fat plaster. All laths are taken out from any formed rebates and the rebates well keyed with a sharp chisel to form undercut keys.

Fixing – general method

Any 25×50 mm battens on the back of the cast are sawn off and their wads trimmed well back below or flush with the strike-offs. Any lumps should also be cleaned from the strike-offs with a saw or plane. The position of the fixing laths in the cast are marked across the face with a chalk line or a pencil and straight-edge. If the cast is to fit into a confined space, it is advisable to check the size with a PINCH ROD and trim if necessary. Excessive trimming that removes the strike-offs will also remove the fixing and leave the edge of the cast weak and unsupported. It is, therefore, of great importance that the cast be accurately set out before casting.

The cast is now ready to be lifted into position. A method of holding it rigidly in place against the fixings is desirable. Several alternatives are available, depending on individual preference and the fixing conditions. Some of the more widely used are T-struts, hangers and cleats (see Fig. 90). The cast may now be fixed.

For casts without additional integral fixing battens the spacing of the fixings should be at 300 mm centres to timber, 450 mm to metal.

Fixing – to timber

Galvanised nails or galvanised, sherardised, plated or brass screws are driven through the strike-off laths into the fixings. Nails are punched home and screws are inserted in countersunk drill holes. Although at the time of writing, most fixing to timber is with galvanised nails, screws may be specified and should always be used when fixing to old work where vibration must be avoided. The overall surface provided by the underside of the fixings needs to be straight, as the nails will pull the cast back against it and reproduce its irregularities on the face. If screws are used, of course, they may be slackened or tightened to regulate the cast, but this is inadvisable as, if the finished work flexes due to timber movement or to pressure on the face, slack screws will cause their stoppings to pop off. Screws are countersunk and nails punched home as fixing proceeds. (See Fig. 91 (a).)

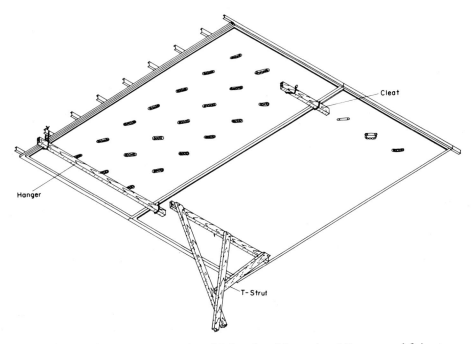

Fig. 90. Plainface (fibrous): hanging plainface for wiring and wadding to metal fixings

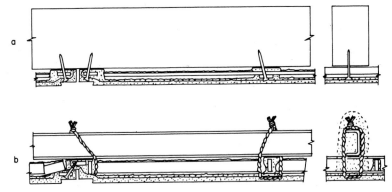

Fig. 91. Plainface (fibrous): fixing (a) with nails to timber, (b) by wire and wad to metal

Fixing – to metal

Ideally, access is needed behind the ceiling in order that the cast may be fixed from the back and such access is usually available because this area is often utilised for electrical, plumbing and air-conditioning services. Two holes are drilled tight against either side of the lath, staggered so that the distance between them is 70 to 80 mm. A groove is now made in the face between them with a chisel, saw or lath-hammer blade for countersinking the fixing wire. This wire is double galvanised tie wire. It is doubled by cutting a length twice as long as that required, folding it in half and twisting it. It is then bent to the form of a square staple, pushed through the holes in the cast from underneath, and twisted over the metal fixing from above.

The final fixing is made by tightening the wire with top cutters. The cast may be strained tight by taking the wire and straining it with the same action as that of withdrawing a nail, then rapidly twisting to take up the resulting slack. The wire is finally cut off, leaving some 10 mm of twist sticking up; a plaster wad is placed over the wire astride the metal fixing and rubbed on to the back of the cast to obtain the best possible adhesion. (Preliminary damping of the back of the cast is unnecessary.) The wad will prevent the wire from untwisting, hold the cast rigidly to the metal, and also help to hold up the cast through adhesion and by gripping round any undercut afforded by the strike-off lath. When the wads have finally set, the wires in the grooves are punched home below the face of the cast.

A cast may also be fixed by drilling through the side of the laths from above, thus not penetrating the face of the cast. With such a fixing the wire will pass straight through the lath rather than, more strongly, across its diagonal, and at best making only half the depth of the lath available for fixing. However, this method is acceptable for securing a cast by its integral fixing battens.

If there is no room to work behind the ceiling, one of two methods will have to be used. In the first the wires are tied from underneath: they are poked through one hole, over the fixing, down through the other hole, tightened, cut off and punched back. Alternatively, hand holes are cut in the cast in order to carry out the wiring and wadding from above and the hand holes later made good with the general stopping.

If enough room exists behind the work for access but there is no way out after the last cast is in posi-tion, all but the last cast is fixed and this latter is then fixed from underneath. (See Fig. 91 (b).)

Fixing – straightening and wadding up

An ideal sequence is for all the casts first to be hung in position by their wires. Two operators then hold a straight edge some 3 m long – or a template, in the case of curved work – against the face of the cast. Each wire is tightened in turn along the length of the straight edge by an operator over the back. At any place where this will pull the cast back, forming a hollow, a small wooden wedge is driven between the fixing and the strike-off so that the cast is forced to touch the rule. This wire may now be tightened to hold the cast rigidly to the fixing. The whole sur-face is straightened in both ways in this manner and every wire then wadded up.

All joints between casts are also wadded; this is done by rubbing a double layer of plaster-soaked canvas over the length of the joints, tucking it in any undercut afforded by the perimeter strike-offs. The back of the casts cannot, of course, be walked on, but one or more scaffold boards may be laid over the metal fixings.

When wadding up over the back of a ceiling it is a great advantage to have passed up from the scaf-fold below, bowls of creamy-gauged plaster con-taining as many canvas wads as the plaster will hold. These should be laid flat to facilitate removal from the bowl. (Wadding up from a bowl can prove a dif-ficult task when one is wriggling around on one's belly, with little or no headroom, on a scaffold board balanced across fixings, among a forest of sharp wires sticking up every 300 mm or so in all directions.)

Alternatively the casts may be hung by integral 25×50 mm battens off fewer but stronger fixings as with BARREL CEILINGS.

Stopping

With plainface this involves making good to all nail/

screw holes or wire grooves and rebated joints. Before gauging, it is advisable to check over the area that is to be made good, in case any major preparation is necessary. In the case of a badly fixed or twisted cast containing local buckles and distortions, more preparation than can be carried out within the picking-up time of the plaster will be necessary. Such areas may need to be cut free, pulled forward or pushed back, re-fixed and stopped in. However, generally the stopping sequence is started by gauging up.

A bowl of creamy plaster is gauged to the consistency of emulsion paint – thinner than that for running or casting. Appropriate lengths of 80 mm jute scrim are put to soak in it. In the five minutes or so that the plaster will take to start picking up, the joint or joints are scraped down with a 300 mm joint rule and any hollow areas cut out and well keyed. The dry surface to be worked on is then thoroughly damped with clean water to prevent excessive suction. When the beginning of set has given the plaster some body the scrim is lifted from the bowl, lightly wrung free of excess plaster and rubbed firmly into the keyed rebate between the two casts, thus wadding them together from the face. The joint rule is then passed along the joint to ensure all the canvas is 'down' and the rest of the plaster in the bowl is used to complete the joint.

The finish is achieved in much the same way as running a mould: further plaster from the bowl is repeatedly applied with a gauging trowel and scraped off with a joint rule. A finish is obtained more quickly if the joint is fed with cheesy plaster by the gauging trowel held in the working hand while the joint rule, held in the other, is scraped over. If necessary, the joint may be scraped with a piece of stiff busk when the plaster has set. A 2.500 litre bowl of creamy plaster should be sufficient to scrim and finish a 3 m or so run of rebated joint.

PLAINFACE, CEMENT AND SAND

See FINISH/RENDERING, EXTERNAL.

PLANK, GYPROC

See BACKGROUND.

PLANTING

See ENRICHMENT, planting.

PLASTER, GYPSUM

Class A

Plaster of Paris, hemihydrate (BS 1191 : Part 1)

It is manufactured from gypsum, $CaSO_4.2H_2O$, which is crushed and heated so that three-quarters of the water content is given off to produce $2CaSO_4.H_2O$. When water is added, the formation of interlocking crystals causes the fluid to set into a solid mass, a process during which heat is given off and the material expands.

Plasters in this class are known as casting plasters. They are available in different grades and strengths. In increasing strength they are: Fine casting plaster, Superfine casting plaster, Teknicast, Helix, Herculite No. 2, Crystacal R, Herculite stone, Crystacast, and Crystacal Alpha K.

Class B

Retarded hemihydrate

Manufactured from a coarser gypsum and with the manufacturer's addition of a retarder. BS 1191 : Part 1 deals with neat plasters.

Finishes

Type b1	Wall finish
Type b2	Board finish
Type b1 and b2	Multi finish

BS 1191 : Part 2 deals with premixed lightweight plasters. Type a are undercoats and type b are finishes.

Type a1	Browning (general purpose undercoat)
	Impact-resistant undercoat
Type a3	Bonding
Type a 1,2,3	Tough coat
Type b1	Lightweight finish

Machine-applied one-coat plaster

A premixed lightweight gypsum plaster with properties that make it suitable for one-coat machine application

Universal one-coat plaster

A premixed lightweight gypsum plaster for hand application.

Hardwall universal undercoat

A premixed lightweight undercoat for hand application.

Renovating plaster

A premixed lightweight undercoat for hand application, for finishing with Renovating finish.

Although Machine applied one-coat, Universal one-coat, Hardwall universal undercoat and Renovating plaster are Class B plasters, they contain lightweight aggregates and therefore do not conform to BS 1191 : Part 1 : Class B. At the same time they are too dense to conform with BS 1191 : Part 2 : Class B.

Renovating finish

For application to Renovating undercoat.

PLASTER, SEASONING

Models for plaster moulds

With the exception of waste moulding (see MOULD, REVERSE, plaster), GREASE is used as the release agent. Grease can be used directly on non-porous surfaces; porous surfaces will need sealing with SHELLAC, and dusting and polishing with FRENCH CHALK. Porous surfaces that must remain unblemished can be treated as for waste moulding, tallowed, or even coated with a polyvinyl alcohol 'plastic envelope' or some similar improvisation.

Although timber requires only greasing in order that the plaster may be released, it is wise to shellac to seal the wood against penetration of water which may cause warping.

Plaster moulds for casting in plaster

With the exception of waste moulding, these are sealed with shellac, dusted with french chalk and greased before each cast.

Plaster moulds for casting in cement mixes

Wet method

Seal with shellac; oil mould before each cast.

Dry method

Seal with shellac and either dust on a layer of french chalk with a muslin bag, or saturate with paraffin and brush on a fresh coat before each cast.

See CASTING IN CEMENT.

Plaster moulds for casting in glass fibre

The manufacturers of the polyester resin used in reinforced plastics recommend the seasoning of dry plaster moulds with cellulose acetate as a sealer. This is then coated with a release agent, polyvinyl alcohol in a water-based or solvent solution; this acts as a plastic envelope and should come away with the cast – any that remains is cleaned from the mould each time.

Coats of wax are also used with these when gloss finishes are required on the cast. Casts can be made to part from plaster moulds, however, by waxing dry plaster with silicone-free wax, by waxing shellacked plaster, and even by coating still-damp moulds of hard casting plaster with petroleum jelly.

The complexity of shape of the mould will affect the ease of release. To obtain the maximum number of casts from a relatively soft plaster mould the first method must be favoured. See also casting in GLASS FIBRE.

PLASTERBOARD

See BACKGROUND.

PLASTERING, MECHANICAL

See MECHANICAL PLASTERING.

PLASTICISER

A liquid aqueous solution of neutralised salt of resin or a blend of similar material with an organic dispersing agent in powder form. When mixed with mortar, it will entrain microscopic bubbles of air within the mix and thus reduce the quantity of water required. This should lessen the risk of the cracking and crazing that is usually associated with rich cement-lime-sand mixes. It will also improve workability. Masonry cement is Portland cement containing the dehydrated form of plasticiser.

Lime

Voids in building SAND amount to a third of the whole volume and must be filled to make the mix workable. To fill the voids with a binder, producing a mix of 1 : 3, would result in too strong a material in many cases. The quantity of binder is therefore reduced and made up by the addition of lime, e.g. 1 part cement: 2 parts lime: 9 parts sand.

PLINTH

In classical architecture a plinth is the lowest square member of the base of a COLUMN. It can also apply to the projecting or stepped member surrounding the base of a building or architectural feature. For practical purposes, the latter will cover the external cement and sand plinth that a plasterer often applies to the outer walls of a building. It usually rises from the ground level, and just clips the top of the horizontal damp-proof course. It may be waterproofed so as to prevent the entry of water or it may be simply decorative. The plain area below the moulded section of an internal skirting is also called the plinth, see also SKIRTING.

PLUMB BOB

The weight on the end of a PLUMB LINE.

PLUMB BOB AND GAUGES

A PLUMB LINE off which work is paralleled by means of gauging templates.

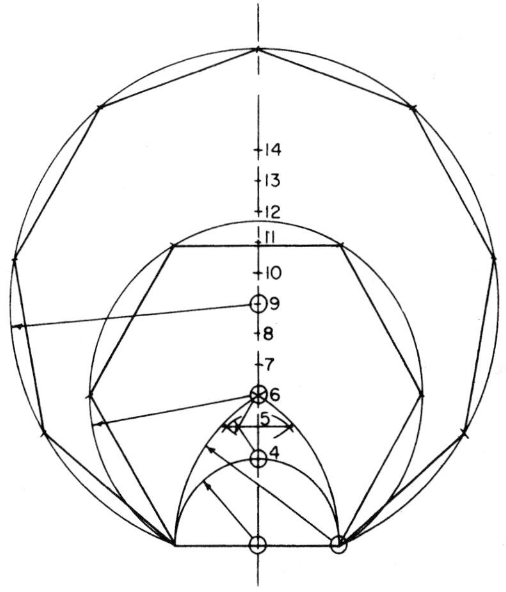

Fig. 92. Polygon (geometry)

PLUMB LINE

A line with a weight attached to the end so that the line hangs plumb.

PLUMBING

See DOTS, plumbing down.

POLYGON

To set out a regular polygon on a given base, see Fig. 92.

The base line is bisected, using its length in the compasses and thus forming equilateral arcs above it. A semi-circle is then drawn, making the base line the diameter. A circle spun from the point where the equilateral arcs meet, passing through the ends of the base line, will be of a size to have the base line stepped round it as chords six times, thus plotting a hexagon.

A circle spun from the point where the semi-circle cuts the vertical bisection line and passing through the ends of the base line will be of a size to have the base line stepped round it as chords four times, thus plotting a square.

Similarly, a circle with its centre the point on the bisection line midway between these at (5) will produce a circle to the base line for a five-sided

195

figure. The distance between points (4) and (5) struck repeatedly along the line from (6) will provide points (7), (8), (9), (10), (11), (12), etc., from which circles to the base line can be drawn to produce a heptagon, octagon, nonagon, decagon, undecagon, duodecagon and so on.

POLYSTYRENE, EXPANDED

See BACKGROUND.

POLYSULPHIDE

See COLD-POUR COMPOUND.

POLYVINYL ACETATE

1 For use as a bonding agent see BACKGROUND.
2 To use the concentrated form neat as an adhesive, porous surfaces should first be primed with a solution diluted 1:5 with water and allowed to dry. Used for bedding enrichment, etc.

POLYVINYL CHLORIDE

The plasterer's PVC is a thermoplastic material made up from vinyl resins specially manufactured for mould making. Various grades of varying flexibility are available; the degree of undercut governs the degree of pliability; the production of good lines and the ability to withstand the pummelling-in required when using stiff casting materials both call for stiffness; the heat generated by some casting materials needs a PVC with a relatively high melting point.

The melting point varies from 120 to 170°C, depending on the grade, and it is advisable to use a thermostatically controlled electric copper for melting, as the material will burn if heated a few degrees above its melting point. Melting may, however, be successfully achieved using an air bath with about 12 mm between two containers. With such an arrangement, the vinyl will need to be constantly stirred to prevent burning – a thermometer must be held constantly in it so that the heat may be adjusted. Observe manufacturer's safety instructions.

Vinyl will reproduce *identically* the texture of the surface of the model.

Use

Models made of clay, plasticine, metal, glass fibre, etc., need no treatment, though prewarming of glass and similar materials may prevent breakage caused by the heat of the PVC. It is best to seal porous surfaces except when the texture of the material is to be reproduced. Plaster and stone may be soaked with water but pouring should take place with no moisture on the surface. A little motor oil may also be brushed into the damp surface but must be fully absorbed. Success on wood may only be possible after sealing with successive coats of two-pack polyurethane varnish. All forms of plaster case containing PVC moulds should be sealed with shellac as porous plaster will draw out the oils contained in the PVC. Molten PVC may be poured on cold PVC without sticking – particularly useful when pouring multi-piece case moulds. PVC is suitable for producing flood, skin, clay-case, run-case and insertion moulds. For manufacture see MOULDS.

PVC is a naturally oily material and therefore needs no seasoning, whether casting in plasters, cements or resins. However, cases are often oiled for pouring so that the case can easily be removed, leaving the PVC on the model; the oiling of large PVC moulds will aid the release of all cement casts and of plaster casts from stiff grades.

A grade of PVC for polyester-resin casting is produced. It has a high melting point of about 170°C and is specially formulated to avoid the fault – often encountered with other grades – of the resin reacting with the PVC to produce a tacky surface on the face of the cast.

POPPING

See FAULTS.

POWER FLOAT

See MECHANICAL PLASTERING.

PRESSED CEMENT CASTING

See CASTING IN CEMENT, dry method.

PRESSED SCREED

See SCREED.

PRICING

See PRODUCTION MANAGEMENT.

PRICKED WORK

See FINISH/RENDERING, EXTERNAL.

PRICKING-UP COAT

See RENDER COAT.

PRODUCTION MANAGEMENT

Whilst mention must be made of contracting and its chain of responsibility in order to indicate the pricing and organisation of work, we do not attempt to deal with the extensive fields of surveying and contract management.

The chain of responsibility

The client is the eventual owner of the building and will pay for the works.

The main contractor is the builder responsible for carrying out the works.

The design is carried out by the architect, who may be working in conjunction with a design consultant, and any one of the sub-contractors may have a separate contract with the architect as a specialist contractor design consultant. However, the chain of responsibility follows the pattern of contracting.

The main contractor, who is usually a building contractor, will let out part of the works for which he has overall responsibility to specialist sub-contractors. These sub-contractors may further sub-let, e.g., a contractor taking all the ceiling works may sub-let the fibrous plaster package. A contractor's responsibility is to the company which has commissioned him to carry out work – the company immediately preceding him in the chain – and all correspondence and contractual proceedings should be with that company.

Any request for variations must be received in writing as an *Instruction* from the main contractor, issued either by him or by the architect, as proof that an instruction affecting costs has taken place.

Pricing

The essentials for pricing are a bill of quantities and tender drawings, consisting of (a) plans and elevations and (b) sections.

The bill of quantities is a list, supplied by the main contractor, of items to be priced. It will have been compiled by taking each item with the quantity of that item in square metres, metres run, number of units, etc., from the drawings issued with the tender. Pricing to a bill of quantities ensures that each tenderer is pricing like quantities.

The bill will have been written by a quantity surveyor in surveyor's language and should be checked against the drawings for two purposes:

1 To check that the take-off is correct in measurement.
2 To ascertain precisely what portions of the drawing are to be priced.

The drawings will enable the pricer to break down the areas of work into individual casts. We can only discuss this process in broad terms because, stock items apart, in general contracting few jobs are ever alike. The factors affecting choice are:

Moulds

It is obvious that moulds cost money to make and, therefore, the fewer used the better. It is advantageous if a mould can be made from which more than one section can be cast or which can be added to or adapted. These considerations may dictate the extent of a cast and the position of joints.

Cast sizes

Within the bounds of practicality (size of bench, transportability, site access), it is generally true that the larger the cast the more economical it is in terms of fewer casts to fix on site, fewer to align and fewer joints to make good. Generally speaking, if five casts of 2.5 m length are being produced in a day from a mould that is considered a 'two-man fill in', then five could still be achieved if the casts were 3 m long.

However, care must be taken not to increase the size of a cast so that disproportionately fewer can be produced in a day and thus efficiency impaired rather than enhanced.

Jointing

The type of joint requiring the least stopping work on site is the one formed where one cast overlaps another, like that of a cornice with a ceiling. Therefore, in an expanse of fibrous plasterwork the obvious place for a joint would be at a piece of moulding or feature where the edge of one cast could be fixed over another. The most labour-intensive joint is one completely in the open; such a joint requires rebating and full scrim-reinforced stopping. See PLAINFACE.

Gridding

The gridwork must be sculpted over the back of fibrous plasterwork to touch the casts at their joints and strike-offs where they need support. Casts that can be made self-supporting with extra cast-in reinforcement will require less grid support work.

If a joint can be made to 'lap' (see PLAINFACE, LAPPED and REBATED JOINT), only one bar is needed to run along the joint to which the first cast is fixed. The second cast is fixed to the first, or through it into the bar. Otherwise, at least two bars would be required, one each side of the joint, or many more bars running at right angles across the joint.

Extra gridwork may be used, however, so that a composite area can be formed with pieces from existing moulds obviating the need to produce a new mould for a complete cast.

The choice is between two basic types of fixing:

1 A more heavily bracketed self-supporting cast hung by wire-and-wad from a grid requiring less supporting members;

2 A more accurately installed grid to line and level with bars at 350 mm centres, to which struck-off casts with flat lath can be screwed with self-tapping screws.

Using (1), less time is required to erect the grid but more is needed to hang, line, level and wad the casts. Using (2), more time is required to erect the grid but less to fix the casts to it. The choice between them for overall economy is made to suit each individual situation, even within different portions of the same job. See GRID, FOR FIBROUS PLASTER-WORK.

It goes without saying that pricing is an area in which mistakes can be costly. A standard form, devised to suit each operation can help to eliminate human error. It will also facilitate future reference to determine the precise methods selected for a job and how its price was reached.

Having decided how the job is to be done, each cast can be put on to the pricing sheet in the form of a sketched section with dimensions. The sketch shows:

1 The shape of the cast.

2 The way in which it is bracketed with the laths making it self-supporting across that span.

3 The position of the two grid members that will touch the back of the cast for fixing.

4 The type of mould, a bed for the soffit, a run moulding for the cheek, a timber rule for the width stop.

5 That the cast is to be fixed with self-tapping screws.

The quantity of this section (item H on the bill) is obtained from the bill and checked against the plan/

elevation to determine the length of the runs and hence the lengths of the casts and the number of metres. We have assumed all lengths at 3 m with no allowance for waste.

As can be seen from the example, it is estimated that, with a two-man team casting at the rate of five per day, the 81 casts will take $32\frac{2}{5}$ man days. It is usual to round up fractions. (Man hours could, of course, be used as a unit instead of man days.) Materials can either be estimated on this form against each particular item, with a percentage handling charge added on, or be expressed as a percentage of the direct labour costs.

When the labour content (man days) is determined it can be multiplied by the total amount a man costs per day (man cost day). This will produce the direct cost.

So that the correct proportion of the company's running costs can be attributed to this particular job, the direct cost is multiplied by the *overhead factor*. This overhead factor has been arrived at by:

1 Predicting how much work the company will do in a year.

2 Estimating the number of production workforce required to do it, and thus their cost.

3 Working out the total overheads the company will incur in that year's operations.

A ratio can then be produced between the total direct cost and the total overheads.

Thus far, the company's outgoings will have been allowed for. It now remains to multiply by the profit factor. The proportion of profit may not be constant; smaller jobs often carry higher profit margins than larger ones. This is because the lower total profit does not give sufficient leeway to allow for unforeseen mishaps and such factors as transport costs, where a lorry taking two casts to a site for one day's work for a man costs the same as it would transporting thirty casts to keep two men going for a week. The final equation would, therefore, be:

$$\text{Man cost days} \times \text{Overhead factor} \times \text{Profit factor} = \text{Price}$$

CLIENT:

CONTRACT:

QUOTE NO.:

DATE:

ITEM	QUANTITY M / M2	UNIT LENGTH	UNITS NO.	UNITS PER DAY	DAYS NO.	MEN PER UNIT	MAN DAYS	TOTAL
H	243 m	3m	81	5	$16\frac{1}{5}$	2	$32\frac{2}{5}$	
Mould								
A		3m	1		$\frac{3}{4}$	1	$\frac{3}{4}$	35
					$\frac{1}{4}$	2	$\frac{1}{2}$	
B		3m	1		1	1	1	
FIX	243 m	3 m	81	6	$13\frac{1}{2}$	2	27	
STOP	Casts	81	81	5	$16\frac{1}{5}$	1	$16\frac{1}{5}$	44
	Mitres	81						
GRID					$5\frac{1}{2}$	2	11	11
								90

When the price has been determined it can be presented to the contractor in one of two ways:

If it is written on the copy of the bill of quantities it must be further broken down. The lump sum for this item must be expressed as the cost per metre run and the cost per mitre. These are entered on the bill in the price column against the item, so that they, of course, total the required amount.

If, on the other hand, the price is submitted in the form of a letter, each item should be qualified – metre run, number of mitres, etc., and the tender drawing number quoted so that precisely what has been quoted for is quite clear. This will enable any future variations to be identified as not having been included in this pricing.

Planning

When a price has been accepted by the contractor a contract will be issued, along with an order for the work. A fresh set of drawings will be supplied, which should be marked *Contract issue* and *For construction*. These drawings should be checked against the tender drawings as they may contain alterations, omissions or additions. These are known as *variations*, as the work varies from that originally quoted. The contractor is informed of the variations and they are priced so that an order for them can be obtained. Sometimes an order is issued before the work has been repriced.

Working drawings

As part of the contract it may be required that drawings are prepared, showing sections of every part of the work and detailed layouts (elevations and plans). These may be required to show (a) the final appearance of the work; (b) the method of securing it in position; (c) the relative position of the plasterwork and its grid to the structure so that the building's services (wiring, heating and ventilation ducting, plumbing, lighting, smoke detectors, sprinkler valves, security systems, etc.) can be arranged in conjunction with it.

These drawings are submitted for approval to the contractor who will send them to the architect or designer for approval; alterations may be required. Indeed, alterations may be required at any time throughout the fulfilling of a contract.

Variation orders or instructions should be obtained from the contractor and a re-price made.

Upon approval of the drawings, they can be put to four other uses:

1 To form a key plan of the work.
2 For manufacture.
3 For fixing the grid.
4 For fixing the units.

Separate drawings showing the grid only may be produced. A single line on a plan will represent each grid member to which the casts are to be fixed. The line will be *tagged* with the same code number as appears on the section drawings, together with its datum height and distance from the building grid lines.

Key plan Four copies of the layout drawings can be turned into key plans by marking them out to indicate each individual cast. These are, respectively, for the use of the office, the workshop, the site fixers, and the site supervisor. Each individual cast needs to be marked with a code and this may be done as follows:

Casts of the same type from one particular mould, regardless of variations, will all bear the same letter. For quick and easy recognition on the plans they can also be colour-coded with a transparent 'Dayglo' felt marker. Each type can be subdivided by adding a number to the letter to denote variations such as rebated edges for joints, lapped edges for joints, extended or shut-down sections, etc. When casts are being made to different lengths they should bear their measurements. On a multi-storey contract, the number of the floor and the cast's position on it may also be marked.

For example, the five pieces of information on the back of a cast might read:

A2 3.200 m Bay 1 3rd floor

However, since there may be several, interchangeable casts bearing this information, the coding is supplied not merely to specify the location of a cast, but so that information about positions on the layout may be relayed from site to workshop or office, provided that all parties have a copy of the key plan.

From these drawings a complete list of the casts can be taken and worksheets prepared to control their production in the workshop. These enable casts to be produced in the correct order to suit site fixing requirements and also enable a check to be kept on the quantity being cast.

Manufacture The full-sized sections are used to indicate how casts are manufactured in the workshop. They already indicate the position of the wood lath reinforcement and bracketing required for casting. The mould can be drawn in round the sections, as it is on the estimating sheet. In addition, running mould profiles are drawn against such portions of the model or mould that are to be produced by running.

Fixing units or grids Copies of the layout drawings, containing the building's GRID LINES and datum lines, are used as fixing drawings. The positions of the datum and grid lines are obtained from the architect's drawings and the position of every feature is set out from these lines. The full-sized sections also indicate the gridwork and are sent to site to be used in conjunction with the layout drawings to set out and fix the gridwork.

Payment

It is usual for payment to be made each month by the contractor in respect of work done. This is done by mutual valuation of the work completed on a predetermined day of the month and usually covers all work completed on site. However, by arrangement, some contractors will agree to off-site valuation. This allows for the valuation of casts stored off-site. In some circumstances, where moulds represent a large initial cost, valuation can be made against their production. Where off-site valuations are made, however, it is usually a requirement that each cast be clearly marked as the property of the client and that a special insurance be taken out for them.

The key plan is instrumental in recording work done. This is fundamental in compiling monthly valuations, as well as monitoring production. On large plans it may be necessary, for monitoring purposes, to add to the office copy five further marks on each cast:

1 To indicate that the cast has been entered on a worksheet for passing to the workshop.
2 To indicate that the cast has been made. (The date of this, should it be needed, will be recorded on the worksheet.)
3 To indicate the cast has been sent to site. (The date will be recorded on the site delivery invoice.)
4 To indicate when the cast has been fixed on site. Here the date may be added to this notation.
5 A further mark (such as a line cancelling the date in (4)) may be added when the cast has been included in a monthly valuation.

Programming

A programme needs to indicate:

1 The period of casting in the workshop, showing which casts are produced, in which order, and by how many men.
2 The period of gridding, showing which areas are gridded, in which order, and by how many men.
3 The period of fixing, showing which areas are fixed, in which order, and by how many men.

The original pricing sheets will contain the number of men and working days allocated to each individual item. From these and similar sheets, covering all variations to date, a bar chart programme can be drawn up. This will be required to indicate the integrated production of casts in the shop, the gridding of areas on site, and the fixing of casts to various areas on site. The timing of the programme has to be done in conjunction with the builder's site programme, not only for the start date or dates, but for the required completion dates. If the given period is shorter than the time estimated more men may have to be used than originally planned.

PROFILE, METAL, FOR RUNNING MOULD

A template made from sheet metal cut and filed to the section of the required moulding. It is held by a running mould in such a way that it may be repeatedly passed through setting plaster in exactly the same place each time, cutting the section from the plaster as it expands. Iron is most commonly used in industry but if the running mould is to be kept and used repeatedly over a period of time, a non-ferrous metal, such as zinc, may be preferred.

Marking the section on to the metal

The following methods can be used:
1 The section is set out and scribed straight on to the sheet metal, using set squares, straight edges, metal compasses and scribes. Dry plaster may be rubbed into the lines to make them more distinct.
2 The same method may be used except that scribing is on to metal that has been coated with engineers' blue.
3 The section may be 'pricked out' on the sheet metal. This is done by placing a full-size drawing of the required section on to the metal and,

with centre punch and tack hammer, pricking all the necessary points. The drawing is then removed and a section drawn through the points.
4 The profile may be set out on a sheet of paper that has previously been stuck to the metal.
5 The section may be set out on a sheet of paper that is subsequently stuck to the metal. (NB: water-based adhesives – PVA, etc. – can cause the paper to stretch and wrinkle, thus distorting the setting out.) Burnt SHELLAC is an excellent instant adhesive. Ordinary thick shellac is painted on to the metal and set alight. The metal is tilted to a near-upright position to ensure rapid, all-over burning. The burning removes all the methylated spirit thinners from the shellac. The heat of the metal will keep the remaining shellac tacky enough for the paper to adhere when pushed firmly down. On cooling, the shellac will set rapidly.

Of the two basic methods – with or without paper – scribing directly on to metal is preferable. It can be filed more accurately, and compasses can be used to cut curves straight through soft metals. The section is cut out with tin shears or a sharp cold chisel on a flat block of metal to within 0.5 mm or so of the line, and the surplus metal removed to the lines by filing and finally brought to a finish with emery cloth or by burnishing with a nail.

PROJECTION

The extent to which a CORNICE or similar moulding protrudes.

PROSCENIUM ARCH

The frame to a stage, the proscenium being the portion of the stage on the audience's side of the arch.

PUGGING

An old term for a material used as a membrane to insulate against sound and heat. It was traditionally a layer of coarse stuff or lime mortar laid on wire fixed to battens between a ceiling and the floor above. See also CEILING, SUSPENDED; HEATING PANEL.

PVA

See Appendix 1, p.279.

PVAc

See POLYVINYL ACETATE.

PVC

See POLYVINYL CHLORIDE.

QUICKLIME

See LIME.

QUIRK

A V-shaped groove, usually less than a right angle. See also MOULDING, SECTIONS.

QUOIN

External angle in a building.

RABBET

See REBATE.

RAIL

See RUNNING CIRCULAR CURVES; SPINNING.

RAKING

In plastering, the term is applied to an inclined feature distorted in some way so that MITRES formed by it are upright or lines contained by it are plumb.

Balusters

In a sloping balustrade the balusters remain plumb while the blocks at top and bottom are tapered.

Blocks – dentil and modillion

On inclined CORNICES the verticals remain plumb. The normally horizontal members slope with the cornice, resulting in a rhombus instead of a rectangle. With a PEDIMENT the blocks on the two sloping members are projected up from those on the horizontal member, thus keeping the horizontal width of the blocks and the distance between them the same.

Mouldings

The sections of these are distorted in order to form orthodox mitres on intersecting with the original (See Fig. 93.). Either may be inclined: horizontal mouldings may be termed raking when they are distorted to pick up with inclined mouldings. To obtain a raking section the projection, as always, is marked out at right angles to the depth. This is horizontal when the cut that obtains the section is vertical (A) but parallel to the slope of the moulding when the cut is perpendicular to the moulding (B).

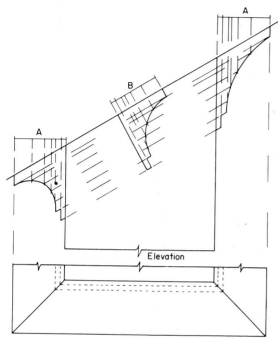

Fig. 93. Raking mould (geometry): to plot the sections of the returns to a pier

In Fig. 94 (a), only the top two members (fillet and cyma recta) are raking on the inclined section. The return at the top is plotted as a raking section. The raking portion (b) is plotted in the same way as for Fig. 94 (a). Ordinates are spun along the moulding to intersect with those dropped from the projection scale marked perpendicularly to the normal to the curve. The return for an open end would be plotted in the same way as for (a). Ordinates are dropped vertically from a horizontal projection scale to intersect with those marked round the face of the curved moulding.

RAMP

The meeting of inclined mouldings or surfaces with horizontal or vertical mouldings or surfaces without the aid of raking sections. Two typical examples of this are the combination of DADO or SKIRTING mouldings running down the inclined line of a staircase with the horizontal mouldings on a landing or on the ground floor.

To obtain form of ramp

Snap lines for horizontal mouldings and snap lines for raking mouldings. Measure equal distances back from the point of intersection, usually 150 mm. Bisect the angle and with either a large compass or radius rod strike a curve from the raking line to the horizontal line. Fix running rules to the marks 150 mm from intersection and fix a greased lath to the edge of the rules and bed it to the curve. Apply Class A plaster to the bent lath and strike off the face of the two running rules. When this has set remove the greased lath, shellac the plaster and use it as a running rule. When several of these are required a metal rib should be cut to the curve. This can then be held in position and the plaster trimmed to this. See Fig. 95 for flush dado moulding.

RAKING MOULD

Figure 93 illustrates the plotting of the two returns to a pier met by a cornice travelling down the soffit of a staircase.

To plot the distorted sections, points are selected round the girth of the regular section (every member and points to delineate the curves). These points are then marked along the face of the moulding, parallel with the run. They are projected perpendicularly up on to the projection, thus forming a scale. This scale is marked as a projection horizontally from the wall that is to receive the raking section. The section is plotted by dropping ordinates from the scale to intersect with the markings along the face of the moulding.

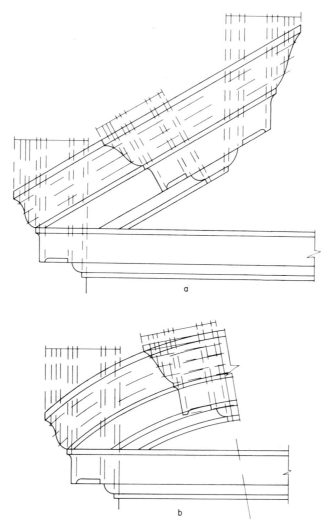

Fig. 94. Raking mould (geometry): (a) to plot the raking sections and returns to an open pediment, (b) to plot the raking section to a curved pediment

Ramped surfaces are inclined walkways or rights of way for vehicles. These are frequently formed in granolithic. The intersection of straight to inclined surfaces will be formed as a curve in the manner described above.

REBATE (RABBET)

1 Rebates in casts

Round the edges of some fibrous plaster casts to

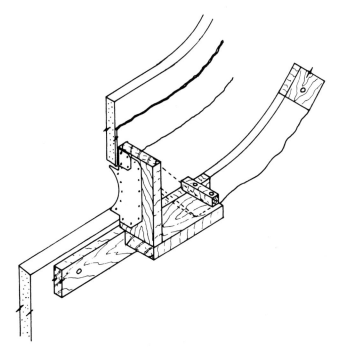

Fig. 95. Ramp (solid): running a flush bead dado moulding

accommodate scrim-reinforced stopping. See PLAINFACE.

2 Rebates on ribs

To receive the lathing when forming a DRUM.

3 Rebates on running moulds

Used chiefly when running *in situ*. A rule is fixed along the back of the slipper. Its function is to rest on top of the running rule, to hold the bottom of the slipper fractionally away from the floating, thus doing away with the need for a slipper screed. (See Figs 35, 36, 112.)

REED

Convex moulding, side by side. The reverse of FLUTE. See also COLUMN.

REFLECTOR COVE

See LIGHTING TROUGH.

REINFORCEMENT

See under particular item to be reinforced.

RELEASE AGENT

A substance applied to a casting surface to prevent the cast from subsequently sticking to the surface. Substances used:
For casting in plaster – GREASE, SOFT SOAP, CLAY WASH.

For casting in cement mixes – MOULD OIL, FRENCH CHALK, paraffin.

For casting in polyester resins – polyvinyl alcohol, wax polish.

For COLD-POUR moulds – polyvinyl alcohol, petroleum jelly.

RENDER COAT

The first coat in all three-coat plastering. It should average 9 mm in thickness and be well KEYED to receive the FLOATING coat.

RENDER STOP

See BEAD, METAL.

RENDERING

See FINISH/RENDERING, EXTERNAL.

REPAIR, FLOOR

See FLOOR LAYING AND ROOF SCREEDING.

REPAIR AND RESTORATION OF OLD LIME PLASTERING

Before reading this entry it is advisable to read the entry on ANTIQUE PLASTERWORK

Repair and restoration of old plasterwork may take many forms, entered here under the method of plastering:

Plain moulded sections that are run with a running mould, either *in situ* or as a plaster reverse mould from which fibrous plaster casts are made.

Enriched work where squeeze moulds are taken from undamaged portions still in place or pieces of plasterwork are removed and taken to a fibrous plaster shop to be moulded in order that casts may be made.

Re-fixing and securing original work that may have become loose due to the plaster key between the wood laths having broken or the laths themselves having rotted away.

The first two forms listed above may fall into two distinct categories:

(a) A close imitation of the finish of the period, i.e. when laying down and dressing a model, faking the slight bending of the supposedly straight lines characterising the workmanship in the old lime stucco (as distinct from the dead straight lines produced in modern gypsum plaster runs).

(b) Merely using old enrichment or enrichment newly carved and modelled to make up fresh models utilising modern methods and materials, workmanship and finish. Such reproduction appears authentic, therefore, only to the layman.

Plain moulded sections

With this, all that is required is an accurate tracing of the section from which the metal profile will be cut. If the whole section of the moulding can be removed without its breaking, all well and good, but with old cornices, etc., the section may be so thin – even going behind the face of the wall and ceiling plaster in places – that this may be practically impossible. Although it is occasionally advocated that a plaster squeeze should be taken, this process is messy and, as the original section was almost certainly run *in situ*, an undercut is often present. A quicker, easier way is to make a saw cut square across the face of the section. A sheet of card or thick paper may then be inserted in the cut and the section traced round on to the paper. For reproduction see RUNNING IN SITU; RUN CAST; or (for replacing the section as a fibrous plaster cast) MOULD, REVERSE. All casts should be fixed to old work with screws to avoid the vibration caused by nailing. The screws should be non-ferrous.

Enriched work – clay squeezes

The simplest use of clay squeezes is in the replacement of small, odd pieces of enrichment that can be cast on site directly from a squeeze and planted in position. This instant repair has the advantage of matching exactly the existing undamaged work right down to the degree of obliteration by decades of paint. A squeeze is taken by pressing modelling clay, plasticine or Vinygel round a selected model, thus forming a mould. The points to observe are:

1 The medium should be worked to the correct consistency, and its surface made absolutely smooth. This may be achieved by pressing it against a sheet of glass.
2 All the surfaces should be french chalked or, in the case of the model, paraffin may be used.
3 The medium should be pressed firmly around the model, care being taken not to cause it to spread and thereby enlarge and distort the squeeze.
4 During the pressing process something flat should be pressed against the back of the squeeze to act as a surface for it to rest on while casting takes place.

In the case of larger squeezes, the clay is applied in small pieces to the face of the model, pressing well to reproduce detail and eliminate joins in the clay. A plaster back may be applied to support the squeeze during the casting operation.

Enriched work – squeeze moulding in the round

Separate areas of clay should be applied to the model so that they will draw from the undercut. The edges should be cleanly cut and paraffined so that one piece does not stick to another. They should be thickened and shaped so that there is no undercut and the plaster back will draw from them. They are peeled from the model and carefully laid into the plaster back. The clay-faced piece mould is now ready for casting. Paraffin brushed lightly over the face with a soft brush will release the clay easily from the plaster cast. However, if it is feared that this will damage fine detail, the clay may be scraped and washed from the face of the cast without the aid of a release agent. Obviously, a replacement of this sort would be easier if the model were removed and taken to a fibrous plaster shop for moulding, but where no shop facilities are available or only one cast is needed, this technique is quite practical.

Enriched work – plaster squeezes as moulds

Taking large plaster squeezes from ceilings can be extremely difficult as this necessitates casting upside-down. If the squeeze is to be used as a mould to produce casts, it will need to have a perfect plaster face. Usually the plasterwork will be already sealed by the coats of paint on it but, if it has been stripped, it will need to be sealed and then greased.

These types of mould can only be taken, of course, if very little or no undercut exists. All undercut will have to be blocked off with clay or filled in with grease. Clay walls can be used to form the edge of the casting area but should be fixed before the surface is greased, in order that they should adhere easily to the ceiling. The easiest part of producing a cast upside-down is probably applying the firstings. It must be accepted that this will be a messy process and everything around should be protected by sheets of polythene.

The firstings needs to be gauged thicker than is usual for casting but is applied in much the same way. It should first be brushed well into all the detail and a layer applied over the whole face. This may be applied with a conventional splash brush or by a combination of splashing and dubbing with a gauging trowel. As this necessitates much working over a period of time with plaster thicker than usual, size should be used. Having applied the firstings, the next stage is crucial. Once the firstings sets and expands, it will split itself away from the greased ceiling and come crashing down. The easiest way round this problem, provided that the moulding is in reasonably low relief, is to have a board of chipboard, plywood or blockboard made ready with the seconds and apply the latter before the firstings fully sets. Enough plaster (that has picked up into its early, cheesy stage) is tipped on to the board to cover it entirely to the required thickness. Depending on the area of the board and the number of competent helpers, this may be up to around 300 mm.

For strengthening thin squeezes of this type, layers of canvas may be put in the plaster, low down, next to the board. Immediately the board has been made ready and while the plaster is as soft as possible, it is lifted up to the firstings and squeezed well on to it. This may be assisted with struts from the floor or a scaffold – necessary to hold the whole squeeze in position until it has finally set. Unlike some more conventional casts, there should be no problem in removing the squeeze from the ceiling as the expansion of such a volume of plaster will release the squeeze from the greased surface.

The writer has taken such squeezes some 750 mm wide and 2 m long formed on the backs of old doors, shellacked and greased them and produced fibrous plaster casts from them, on site, to repair cornice and ceiling ornament.

If a board is not made use of (due to the work being in deep relief or on curved backgrounds, e.g.

pendentives, barrel ceilings) the following system may be applied:

The firstings should be applied as above but to a greater thickness so that when set it will be self-supporting over areas 300 to 400 mm wide. While the firstings is being applied, previously cut 25×50 mm battens are made ready by having juicy ropes of 300 mm wide canvas soaked in plaster placed along them. As the firstings picks up, the battens should be lifted from the scaffold and pressed on to the high points some 300 to 400 mm apart, strutting each one from the floor or the scaffold.

If the work is too high to be strutted from the floor, the scaffold must be specially braced off the walls so that it does not move. These struts should be numerous enough and positioned in such a way that they will support the whole squeeze, bearing in mind that it will be much heavier than a conventional cast of a comparable size. These struts and bearers should hold the firstings to the ceiling, and as the squeeze hardens further, it is completed by throwing up soaked canvas wads to cover the whole area, including the battens. Two layers of canvas is the required minimum. Cross-brackets of 6 mm laths or 25×25 mm battens should be included in this operation. In order to have the cast strengthened across its width, it may be desirable to wad 25×50 mm battens at right angles across the original ones.

The easiest way to apply the wads is to put as many as possible in a bucket of neat, creamy plaster of Paris and leave them to soak until the plaster starts to pick up. In this state the wads can be made to stick more easily.

Enriched work – mouldings

This will involve a combination of the techniques previously described. The repair may be effected by (a) running the moulding containing beds for the enrichment *in situ* or on the bench and dressing it *in situ*, or (b) running the model on the bench with beds for the enrichment, dressing it, moulding it and making the necessary casts from the mould. In each case a positive profile is needed. Alternatively, a reverse mould may be run so that the casts taken from it contain beds that are dressed *in situ*. For method see RUNNING IN SITU and MODEL MAKING.

After taking a profile the second operation is to take down specimens of all the enrichment and prepare them for moulding. As stated previously, it may be necessary to mould paint and all – so that the repair matches the existing work. If, however, the repair involves replacing all the plasterwork or the work remaining is to have the paint removed, the enrichment for moulding must also be cleaned of paint. Indeed, it is good policy to remove paint in order that the reproduction matches the original work and have it blinded up with paint to the same degree as that of the existing finish in order that, if the paint should be stripped at a future date, all the work will match.

From the time of the earliest paintwork in this country two main types of paint were always available for decoration:

1 The traditional *oil paint* made up from linseed oil, lead pigment and turpentine with driers.
2 *Distemper*: a composition of common whiting bound with either glue size or casein. This was the finish most commonly used for internal plaster surfaces. The variety consisting of a body of whiting (powdered chalk) bound together with size water was often left white but might contain a pigment. A coating of this form of paint can set extremely hard due to the size. 'Washable' distemper was made by mixing casein (milk protein in powder form) with whiting and lime. Casein is not soluble in water but is made soluble by the addition of an alkali – in this case, lime. When applied as a paint film, the action of the lime ceases when the water evaporates and the casein reverts to its hard, insoluble form. This distemper has improved

qualities of adhesion and when hard is 'water repellent' in that it withstands light sponging.

Both types of distemper will not be substantially affected by modern paint removers but can be readily softened and dissolved by hot water. During redecoration in Victorian times the water distemper was often washed off. It readily came off completely from the plain moulded sections but remained with the enrichment and filled it. If at this stage a coat of oil paint was applied it would, of course, be next to the plain sections and its removal may not be necessary before obtaining a profile for the running mould. But when it is encountered underneath layers of distemper subsequently applied, it must be removed from the enrichment so that the distemper build-up underneath it may be washed off.

The pastel colours used by the Adam brothers were mainly oil-bound paints rubbed down afterwards to flatten the gloss. If an oil-bound paint was applied over coats of distemper it may still be removed by boiling, otherwise paint remover may have to be used. Paint from the later Victorian mouldings may be removed by working it under a hot-water tap, but painted earlier work will need to be simmered gently in a tray of boiling water, until the paint softens and may be peeled off in layers. Enrichment in Georgian plastering may have become completely obliterated and the gradual removal of paint brings to light the skill of the craftsman – a fascinating and exciting experience.

Securing old, plain ceilings – sound lathwork but failing key

In some cases, when the key has failed, a ceiling will be loose but sagging by no more than a few millimetres. In such a case it may be re-fixed to the lathwork, and even the joists, from underneath by plugging with canvas wads. The spacing of these wads will depend on the condition of the ceiling: somewhere between a 300 mm square grid and a 600 mm square grid should suffice. Holes of about 30 to 40 mm in diameter are cut through the plasterwork. These should reveal gaps between the wood laths.

Each hole should be dovetailed so that the plug will hold up the old work. It is advisable to paint a dilute coat of PVA bonding agent round the inside of the hole in order to reduce suction, penetrate and stabilise the sand–lime undercoat, and seal the old work. Sealing is necessary as otherwise the water used will bring a stain on to the surface from the old sand–lime mix. Small canvas wads are soaked in a bowl of creamy plaster of Paris and, as they stiffen, one is pushed into each hole so that part of it passes between the laths while the rest fills the hole to within about a millimetre of the surface. This may then be stopped in with the remainder of the same gauge.

If extra filling is required the joists should be located, and similar plug holes cut along underneath them. Non-ferrous screws may then be driven into the joists at an angle to afford the best possible key for the plugs. These should be sunk just below the surface of the ceiling and the work sealed, plugged and finished as before. Cracks in old work should be cut out, removing all loose material, and brushed thoroughly before sealing with PVA and making good. It is obvious that it is not wise to hammer on such a ceiling as the vibration may cause further damage.

If pieces of the ceiling have actually come down, the plaster should be cut smooth round the edge of the hole. Care should be taken to splay the cut outwards so that the new patch will act as a dovetailed plug. The edge of the old plaster should be sealed with PVA. When making patches good the absence of suction where the new work meets the old greatly assists the production of a flush, undetectable join. If necessary, canvas plugs may be put against the sides 300 mm or so apart.

To make patches good in old sound work, one or two layers of plasterboard may be nailed on top of the laths and the area finished with a one-coat board finish plaster.

However, fragile ceilings must not be banged and this precludes the nailing of plasterboard. In this case, the thickness of the old plaster may be brought out quite easily with lightweight metal-lathing plaster. A type of artificial bonding may be produced by gauging plaster of Paris, containing size, to a creamy consistency and stirring in coarse vermiculite. This mix has the advantage of a controllable setting time.

Securing old ceilings – jacking up

This operation will vary according to the complexity of the ceiling ornament. If the ceiling is plain or contains low relief it should be covered with carefully cut and positioned boards that are strutted from the floor or, if the ceiling is too high, from a scaffold. A cushioning layer of paper and expanded polystyrene should be placed between the boards and ceiling. Scaffolding, if used, should be braced against the walls to prevent movement.

Securing old ceilings – fixing

When all the affected plasterwork is supported, any defective joists may be removed and replaced. If the only fault is that the key is broken the old wood laths will still have to be removed. This is effected by sawing through the laths tight against both sides of every joist and poking them through. In the event of the laths having rotted, the back of the ceiling will just need to be cleaned by brushing. In this case, provided the backing coat is strong enough, the key will form an excellent mechanical key for the new plaster.

The use of the form of strutting described is only feasible where the backing coat is sufficiently strong. If a ceiling is worth preserving but the backing coats are disintegrating, the old finishing coat must be supported by full plaster squeezes.

All rubble, pieces of rotted lath and loose friable material must be removed from the back of the ceiling – finishing off with a vacuum cleaner. The ceiling should be pushed back in position, care being taken to avoid all unnecessary vibration. This is best done by levering still-sloping struts more upright and by using folding wedges to push up vertical struts.

Before the use of PVA adhesives and bonding agents a thin coat of shellac was applied to the back of the ceiling to kill some of the suction. Nowadays, a solution of PVA will do this and also improve adhesion. A cast can now be made, bit by bit, over the back, treating the old ceiling as set firstings. Plaster of Paris, gauged with size, should be brushed well into the surface. The back should be covered with the conventional two layers of canvas. Cross-brackets may be incorporated at right angles between the joists, depending on the span, weight of the ceiling, etc. The casting is fixed by wads to the joists during this operation. To conform with modern methods of fixing by securing casts with metal, e.g. wire and wad or nails/screws, chicken wire may be incorporated with the canvas and fixed over nails splayed in timber joists or over metal bearers.

Squeezes as supports

This technique is necessary where the depth of the moulded work or the contours of the ceiling require more support than that afforded by flat boards, or in cases where better support is required as the old coarse stuff backing is to be removed.

Areas of old plasterwork – anything up to whole ceilings – may need to be supported:

1 While they are re-fixed.
2 To push them back into their original positions from which they have sagged.
3 While the plasterwork is removed in sections to

be re-fixed on a new site or transported to a fibrous plaster shop where its paint is removed and it is subsequently used as a model.

As the squeeze is not to be used as a mould, the quality of the face is not a factor in its production. The problem of the first plaster applied coming away from the ceiling as it expands still has to be coped with. It is necessary, therefore, to get everything completely prepared beforehand and laid out ready for easy use. That is to say, 25 × 50 mm cross-members and the struts to hold them against the ceiling, canvas laps, and all other minor timber brackets, such as 6 mm lath, 25 × 25 mm, or 25 × 50 mm battens.

In all three cases mentioned, the ceiling is covered with squeezes of a manageable size. In cases (1) and (2), the necessary squeezes may consist of anything ranging from mere rope-covered battens to a full plaster cast over the whole area. The latter, however, will have to be much stronger and self-supporting as it is to act as a packing case for the old work once it has been taken down.

Squeezes as supports – producing the squeeze

This may be carried out by bedding a plaster-covered board on to the ceiling as described earlier, but size and shape, dictated by the repeats of the ornament, may make this impracticable; so one which more closely resembles a conventional cast will be employed.

To start with, all undercuts should be blocked out with clay. If it is preferred that the plasterwork should not be greased, wet newspaper will follow the contours of the moulding and can sometimes be made to serve as a separating membrane. Clay walls may be used to form the perimeter of the squeeze. The cast is formed structurally in reverse order and not necessarily in one continuous operation. To keep the cast against the ceiling, the bearers are fixed first across its face, and one only need be applied in each operation. The operation proceeds as above for enriched work – plaster squeezes as moulds. In the cases of (a) and (b) the procedure is as for securing old ceilings.

Taking down for re-fixing later

When the squeeze has set, the old ceiling should be cut through tight against the edges. If a whole ceiling is being dealt with all the squeezes may be made, leaving the 20 mm wide clay walls round them in position. A keyhole saw may now be used to cut through the old ceiling round the perimeter of each cast via the clay. One at a time, the casts are carefully lowered down to the floor or scaffold directly beneath them. Before lowering, the ceiling has to be released from its background.

The procedure is now the same as in securing old ceilings, except that the casting over the back is formed in the same way as the back of a conventional cast with cross-brackets or bearers positioned on edge so as to provide a fixing for wire and wads. During re-fixing, the wires will be placed round or through the cross-brackets in such a way that the face is not penetrated. When the back is completed, wads are positioned every 300 mm round the perimeter, fixing the new back of the old ceiling to the squeeze mould support. The sandwich-like construction will act as a packing case, affording protection to the old work.

On re-fixing, the squeeze may be removed immediately before fixing, or left on the cast and the whole strutted in position for wiring. The casts are then fixed in the conventional way, the squeezes removed from their faces and all the joints stopped in, sealing the old edges with PVA.

Re-fixing and reinforcing unsafe fibrous plaster

Fibrous plastering was patented as long ago as 1856, and laths and canvas contained by old casts may have rotted where bad conditions exist behind the work. When reinforcing the fixing of such work, it may be that there is nothing left in the cast to fix to and the casts are, therefore, secured in the same way as the older lime plastering on lathwork. It is advisable to make use of any existing mechanical key, such as that afforded by the material covering on-edge lath and the struck-off flat laths. The back is thoroughly cleaned, coated with a sealer/bonder, and all necessary casting carried out over the back, including fixing to bearers.

RESIN

A natural substance found in fir and pine trees. Extracted as a sticky substance, allowed to harden and ground.

RESIN, SYNTHETIC

See GLASS FIBRE.

RESTORATION

See REPAIR AND RESTORATION.

RETARDER

An ADDITIVE that slows down the set, hardening or curing of a mix.

RETICULATION

See FINISH/RENDERING, EXTERNAL.

RETURNED END

A moulding returning on itself and within its own projection, forming an external mitre.

REVEAL

The surface, usually at right angles to the main wall face, at the side of an opening – door or window.

REVERSE MOULD

Any mould from which a cast is taken. See MOULD.

RIBBED CEILING

See CEILING, RIBBED.

RIBBED VAULTING

See VAULTING.

RIB

A rib consists of a curved plaster rule, containing timber and canvas reinforcement, supported by being joined to a base rule by a series of struts. Ribs are used as running rules off which peg moulds are run for both bench and *in situ* work, rules to form pressed screeds for *in situ* work, rebate-carrying templates off which DRUMS are lathed out and ruled in, and portable casting rules for curved plainface.

Casting

First the shape of the curve is set out on the bench and a suitable curved RULE provided, containing the appropriate rebates, etc., to suit the purpose to which the rib is to be put. Straight casting rules are then fixed to form the rib's base and ends. The whole is greased, reinforcing materials cut, and the casting carried out in the following way:

Firstings is applied to the inner face of the curved rule and the adjacent bench, forming a 100 mm or so wide band round the inside of the rule. Canvas is applied round the curve in the normal way and brushed in with seconds. A rope is placed against the curved rule. Laps are soaked and placed flat against the base casting rule where each strut is to be joined. Two layers of 6 mm lath are positioned flat on the rope and laid with break joints for strength. These laths are, of course, placed to receive any nails during use. The base rule and all 25×50 mm battens are put in position and the canvas turned in. The laps will form wads, join the struts to the base rule and the perimeter canvas will complete the curved rule and join it to the struts. The curved rule is then struck off, plain or with a rebate gauge. For use see DRUMS.

See also CEILING, RIBBED; VAULTING, RIBBED.

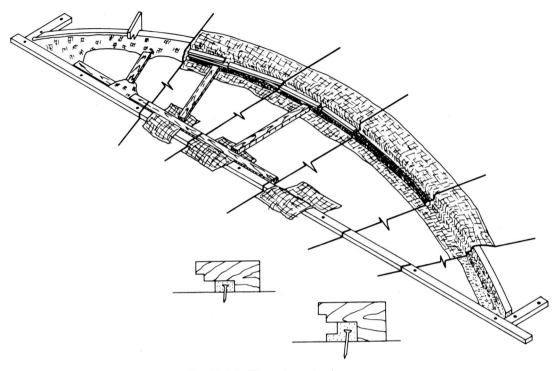

Fig. 96. Rib (fibrous): casting in stages

RISER

1 Upstand. An upright section of a reverse mould, usually loose as a split piece. See also BEAM CASE.
2 See STAIRCASE WORK IN GRANITE AND CEMENT.

RIVEN LATH

See BACKGROUND.

ROD, STOREY

See STAIRCASE WORK IN GRANITE AND CEMENT.

ROMAN CEMENT

See CEMENT, ROMAN.

ROPE

See CASTING IN FIBROUS PLASTER, general method.

ROSIN

See RESIN.

ROUGHCAST

See FINISH/RENDERING, EXTERNAL.

ROUND

Stiff gauge of plaster.

RULE, BACKED

For cornice work on external angles (Fig. 97 (a)). A piece of board that acts as the continuation of the slipper's screed is fixed behind the running rule to provide a surface on which the slipper may continue to run. This allows the running mould to run past the angle so that a tight mitre can be cut.

RULE, BURIED

See RULE, ECCENTRIC.

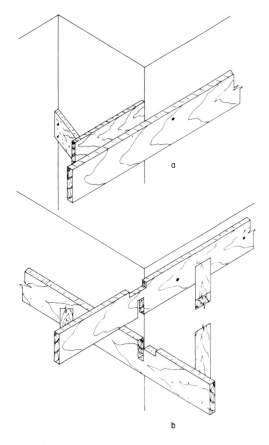

Fig. 97. (a) backed rule, (b) notched rule

RULE, CURVED

Curved rules can be considered in the light of the degree of accuracy required of them. Poor or relatively inaccurate lines may suffice for casting shaped PLAINFACE where 10 to 20 mm tolerance is allowed for in the joints, and for casting RIBS that will be lathed out to form a core for large runs, etc. Such rules may be provided by bending a lath on edge to a line. Other lines need to be extremely accurate. Examples are rules for a peg mould, risers for LIGHTING TROUGHS, and ribs for DRUMS, peg moulds and forming pressed SCREEDS. These may be produced as follows:

Plaster band

Either spun or run with a trammel, usually a fixture and cannot be moved.

Timber rule

Bent to curve. While rule itself is portable, the curve has to be set out afresh each time.

Plaster template

Cast from one of the above, portable.

Plaster riser

May be needed to form an upstand to a plainface for forming floats, lighting troughs, etc. If it is to be moved around a lot, a cast fibre riser (as opposed to a run solid one) will be needed.

Entasis rule

See COLUMN, diminished.

RULE, ECCENTRIC

On flat surfaces

This method of running keeps the stock of the running mould constantly at a normal to the curve. The eccentric rule is so named because it does not follow parallel to the curve being run and, as can be seen in the shape invented to demonstrate this in Fig. 98, at a point where a straight line changes to a curve the rule will kick violently. The rule, always in conjunction with another rule, can be used to run any changing curve. The half-slippered running mould fits between the rules and is forced against both of them by a continuous clockwise-turning pressure. The second rule may be fixed at one of three positions (Fig. 99 (a)): (i) as a nib rule; (ii) against the inside of the first member of the profile, the stock being made flush with the profile at this point to act as the bearer – in which case the rule, being covered by the moulding being run, is known as a buried rule; (iii) against the inside front of the slipper, the stock being suitably notched to allow this. As can be seen, when a nib rule is used the two bearing points make a much greater angle with the two rules, increasing the likelihood of the mould's jamming and running less smoothly (Fig. 99 (b)).

Setting out

(i) The line of the moulding is set out on the running surface and, when it is being used, a buried rule is

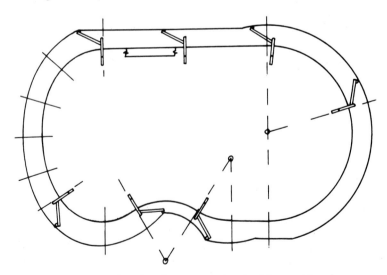

Fig. 98. Running changing curves: the plotted eccentric rule

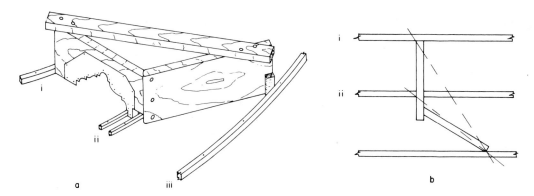

Fig. 99. Running changing curves: with an eccentric rule

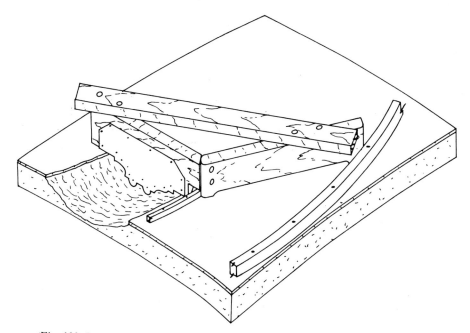

Fig. 100. Running changing curves: on curved backgrounds with an eccentric rule

fixed to this line; (ii) normals to the curve are marked on the running surface by geometrical methods or with a radial square, if this is thought to be sufficiently accurate; (iii) the running mould is offered on to each normal in turn and the position of the back of the end of the slipper marked on the surface. These marks plot the path of the eccentric rule and it is fixed to them. It is as well to fix a temporary strut to hold the slipper at an angle to the stock in order that the best angle may be found. Where a curve changes from concave to convex a much greater angle is needed to prevent the slipper from fouling the rules. When a nib or slipper rule is to be used, its position is marked along with that of the eccentric rule. Where the path of the rule includes tight curves and sudden kinks, any timber rules may more easily be bent if soaked in water; metal rods or square sections, backed up by a continuous wad, may be used, or the curve may be cut from plywood, etc.

On curved surfaces

Running moulds for these conditions are of the same half-slippered type as on flat surfaces (see above) but are cut away to form the three-legged mould needed to bear on curved surfaces (Fig. 100). The stock is feathered so that the profile is kept in contact with the surface as the mould tilts, and the stock is sometimes fixed out of the perpendicular (as with all three-legged moulds to run on curved backgrounds. See SPINNING ON CURVED SURFACES.)

Setting out

The rules are fixed to the running mould held at normals to the curve, as on flat surfaces.

For uses see NICHE, BEAM CASE.

RULE, NIB

See ANGLE, EXTERNAL RUN, internal plastering; CORNICE, SOLID, EXTERNAL; RULE; ECCENTRIC.

RULE, NOTCHED

For cornice work on external angles (Fig. 97). A rebate on the slipper of the running mould will obvi-
ate the need to back the running rule, permitting both rules (suitably notched) to clear the rebate. The running mould, able to run clear of the angle from both directions, can be used to run up both lengths of moulding and form the external mitre in one operation.

RULE, RUNNING

Usually a timber batten, straight or curved, fixed to a background, for a running MOULD to slide along. It is usually nailed to the background but may be wadded. Alternatives are metal (for quick curves), cut templates or cast RIBS.

RULING IN

The operation of ruling newly applied plaster to a flat and level surface to either screeds or grounds with the appropriate RULE. Or, as with a setting coat, to an average thickness as prescribed.

RUN CASTS

Run casting is a combination of casting and running to produce a plaster section, either as a part of a mould or as a substitute for a cast. For running and casting see RUNNING ON A BENCH and CASTING IN FIBROUS PLASTER.

Run casting is an appropriate method in the following circumstances:
1 When a moulding is to be made on site but is run down, containing reinforcement for fixing purposes, rather than run *in situ*.
2 Where only one or two mouldings are required the operation of producing a plaster mould is eliminated.

3   To avoid a highly complex piece mould, due to undercut in the section, when perhaps only one or two casts are required.
4   Where reverse moulds are to be made up of large sections, material will be saved by making them hollow – and so lighter – and they may be provided with the same reinforcement as the conventional cast.
5   Where a section is moulded on both back and front.

Reinforcement

The method and quantity of reinforcement will vary according to the purpose to which the run cast is to be put. If the section is to be part of a reverse mould, it may only require a couple of layers of canvas; a small section for fixing may need canvas and a single longitudinal lath. If, however, a section is wide, as in the instance of a LIGHTING TROUGH, it will require reinforcing as for a normal cast or, as run casts are many times heavier than conventional casts, relatively more reinforcement may be needed. This may be achieved by placing the wooding closer together or incorporating metal sections.

General method

(a) The simpler form of run cast is merely a run containing reinforcement where canvas on the back does not matter. Sections that fall into this category are mouldings to be fixed, either solid or made hollow by running them over a CORE, or parts of reverse moulds, such as BEAM-CASE cheeks (Fig. 101) and NICHE shafts and heads. In these cases, plaster of Paris – containing no size – is gauged as for running in the normal way. A quantity of plaster is taken from the container, and the canvas soaked in it so that no strands of canvas will contaminate the plaster for the run. The canvas is then stretched along the run, a little extra plaster being placed

under it first if required. Wood laths may now be positioned, the canvas turned in, the running mould passed over to check that no canvas is touching the profile, and the section run up in the usual way.
(b) Where a fair face is needed all the way round the run cast, the underneath is formed by a reverse mould. The upper metal profile is first cut and filed and horsed into a running mould. The lower profile is then filed and fixed in position over the upper one by nailing through it into the stock. In places where the lower profile extends too far below the upper it will not be stiffened sufficiently by the stock and will need to be backed up by a plaster wad. A metal profile is securely fixed to a backing wad by passing nails through previously made holes in the metal.

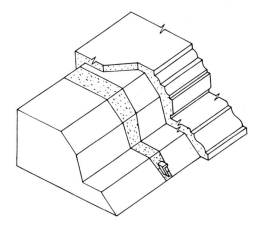

Fig. 101. Run cast

The wad is then pushed firmly round the nails which are put in at an angle to prevent the metal being pulled away from the wad. The core or mould is then run off and prepared for casting by shellacking and greasing in the normal way. The lower profile may now be removed in preparation for the run. If the design of the reverse mould requires the upper profile to touch the mould or core at the strike-offs, etc., the running mould should be repeatedly passed over the section until the plaster stops expanding. This will cut down the strike-offs and so permit its easy passage during run casting.

The run cast may be produced by one- or two-gauge work. The mould is coated with firstings in the normal way and then struck off. The canvas is applied and brushed in. The wood lath is pasted and bedded in the plaster with or without the use of ropes. The canvas is then turned in and pasted. The strike-offs, bench and running rule are then cleaned and the running mould passed over to ensure that the profile clears the back of the cast. If the bulk of the running is carried out with the same plaster that has been used for the casting, the usual difficulties of running with sized plaster will be encountered (see CORE). Therefore, it is sometimes preferred to gauge only enough sized plaster to complete the casting operation and then to carry on running with some freshly gauged, neat plaster, ensuring that the first of this is applied to the cast before the latter has set. In this way a perfect union is obtained. Nevertheless, the cast should be correctly sized and setting at this stage so that the hardening run is not floating around on a soft base.

RUNNING CHANGING CURVES

On flat surfaces

Eccentric rule

This method keeps the stock of the running mould at a normal to the curve throughout the entire run, thereby avoiding reduction of width in the moulding, and the consequent distortion. See also RULE, ECCENTRIC.

Peg mould

Although it is possible to place a peg mould against a rule fixed to a changing curve and run it round, it is impossible to offer the running mould on to the line of the moulding and fix a rule to its pegs. As can be seen from the shape chosen to illustrate this in Fig. 102(a), the leading edge of the slipper will follow a parallel path to the line of the moulding longer than that of the back. A rule fixed midway between these two paths will produce the double bend in the moulding, as shown. However, this

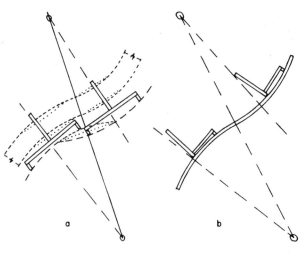

Fig. 102. Peg moulds

method may be acceptable where the running mould is small in comparison with the curve, e.g. when running large ellipses.

Half-slippered peg mould

The line of the rule is set out parallel to the line of the moulding and the stock – bearing against the rule – will run accurately to the moulding line except for the distortion caused by the end of the slipper that rests against the same running rule forcing the stock out of a normal to the curve and so reducing the width of the moulding. The acceptability of this method depends on the degree of distortion brought about by individual curves (Fig. 102(b)).

On curved surfaces

Eccentric rule
See RULE, ECCENTRIC.

Peg mould

The setting out and fixing of the rule is the same as on a flat surface. A half-slippered, three-legged

peg mould with a chamfered stock will be required. See RUNNING CIRCULAR CURVES – ON CURVED SURFACES.

RUNNING CIRCULAR CURVES

On flat surfaces

Although spinning obviously produces the best line, other methods exist to run circular mouldings and these are chosen mainly when the use of a GIGSTICK would be impracticable.

Peg mould

As concentric circles are parallel, the stock will remain a normal to the curve as the running mould runs against a curved rule fixed inside or outside the moulding (Fig. 103). The ends of the slipper will rest on a rule fixed outside the moulding and so requiring no pegs. A mould running on a rule fixed inside the curve, however, will require a peg on each end of the slipper to bear on the rule. In both cases the running mould is offered on to the line of the

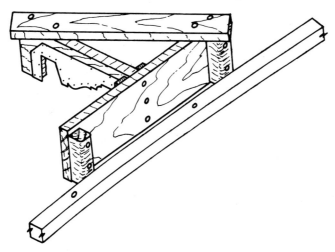

Fig. 103. Running changing curves: on flat backgrounds with a peg mould

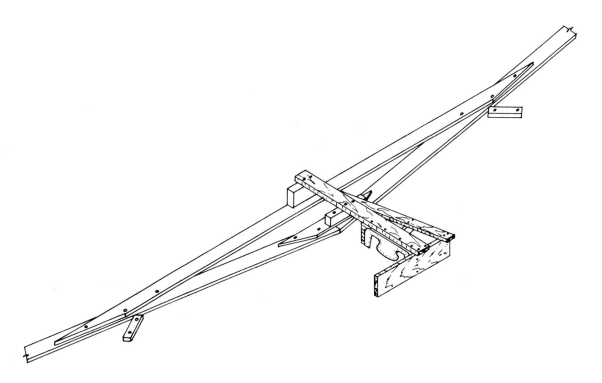

Fig. 104. Running circular curves: the trammel

moulding with its stock a normal to the curve, and a second line plotted to touch the back of the slipper. See also under specific item to be run.

Trammel

A suitably braced template can be constructed to the angle of the apex of a triangle made between the rise and the span of an arc (Fig. 104). The apex will trace the arc when the template is slid round against the two pins at each end of the span. The pins may be timber blocks or substantial screws or nails passing through metal plates to prevent any working loose. The true curve is set out as described, the running mould is offered on and the two pins and the rise moved inwards a suitable distance with a radial square. The angle template is then altered to

fit the new angle. The running mould is fixed to the apex of the template so that the profile fits the original line, with the stock a normal to the curve.

The limitations of this method are that the template must have sufficient room to travel half the span of the arc past each end of the run, and that the profile does not remain a normal to the curve.

On curved surfaces

On barrel vaulting or elliptical domes

This method may be chosen instead of that using the hinged GIGSTICK. The running rule is fixed to the slipper line of the running mould, with the stock held at a normal to the curve that has been spun from a centre point. Alternatively, when it is desired

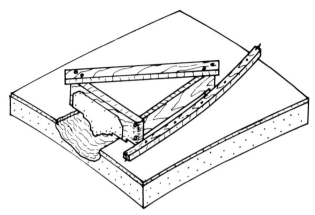

Fig. 105. Running changing curves: on curved backgrounds

that the circle appears from underneath to be a true circle, the line to which the moulding is run is plotted down on to the surface of the drum in the direction required from a circular template. A half-slippered three-legged peg mould will be used (Fig. 105).

The end of the stock and the far end of the slipper will touch the curved running rule, both ends of the stock and the far end of the slipper being the three places bearing on the face of the drum. The stock is fixed at less than a right angle to the slipper in order that, when the mould is against the rule, the stock is at a normal to the curve. The stock is tilted out of perpendicular to the running surface to halve the distortion in depth as when SPINNING ON CURVED SURFACES; also the stock is feathered in order that the profile remains in contact with the surface as the stock rolls. (A circle drawn on these backgrounds will be slightly elliptical and you are, in fact, running a changing curve with a peg mould. Nevertheless, the figure is so near a circle that the distortion in the width of the moulding is acceptable. The alternative is to use an eccentric rule.)

The running rule may be of either metal or wood. On tight curves a metal rod or square section can be bent to a perfect line and will provide an excellent surface to run against. Shoes may be placed on the

running mould's bearing surfaces. A continuous wad on the outside of the rule will hold it rigidly in position, a few nails being toshed into the drum. The size of a timber rule will vary according to the quickness of the curve. It may more easily be bent to a quick curve when it has been soaked in a LATH TANK. When the fixing pins are likely to split it, the rule may be drilled. See also NICHE, curved backgrounds; RULE, ECCENTRIC.

Solid

See ARCH; CORNICE; RAMP; VAULTING, RIBBED.

RUNNING, DIMINISHED

Several methods exist for running a tapering moulding but each produces its own kind of distortion. Basically the effect is achieved either by twisting the entire stock at an angle to the run (and thereby reducing its width) or by folding the stock at the centre by means of hinges.

Diminishing a run in width only

Suitable only for running on flat surfaces and circular domes (Fig. 106). With this method a specially horsed running mould is forced out against two running rules by a continuous clockwise twisting pressure. The suitably notched stock fits between the rules and bears on the inside.

This hinge mould may be run on flat surfaces and over semi-circular domes, as Fig. 107 shows. With this method hinges are used to allow the stock to turn at an angle to the run. They join the ends of

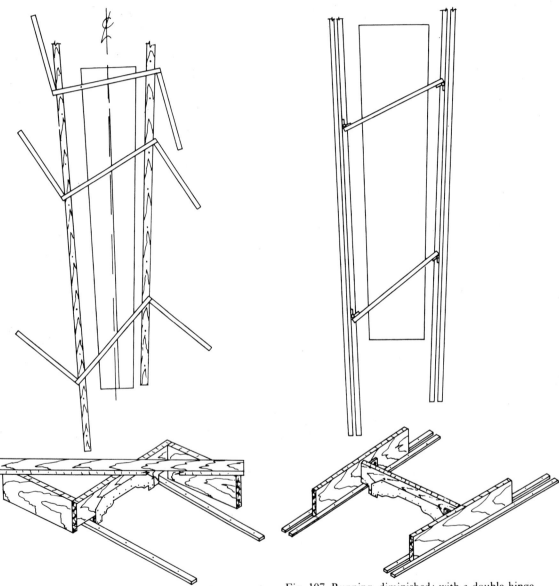

Fig. 106. Running, diminished: with a twisting mould

Fig. 107. Running, diminished: with a double-hinge mould

the stock to the centres of the slippers, one on the back of the stock and the other at the front. Each slipper runs between a pair of rules, the stock being suitably notched over the inside rules to allow this.

With both methods the width of the diminished moulding is marked on the running surface. The profile of the running mould is offered on to the setting out every 100 mm or so along its length, twisting the stock to make the profile fit the setting out. The position of the two parts of the running mould that will bear against the running rules is marked at each point, and the running rules then fixed to these marks.

The methods are suitable for running long, slight diminishes but produce too great a distortion in short, quick tapers. The distortion is shown in Fig. 108. All lines are pulled off centre towards the trailing edge of the stock.

The triple-hinge mould of Fig. 109 may be used on flat surfaces, over a semi-circular dome or on barrel vaulting. The stock is hinged at the centre front, and both ends hinged to the slippers at the back, thus allowing the stock to fold as the slippers come together. Running is greatly assisted by a board being fixed over the mould to the left-hand slipper only. This will hold the stock down and

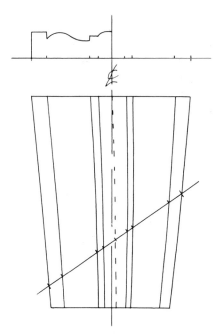

Fig. 108. Running, diminished: distortion caused by the methods of Figs. 106 and 107

prevent it from lifting due to the play in the hinges. It can also be made to carry a rule to act as a stop for the right-hand slipper which is forced against it as it slides in towards the left-hand one. This will

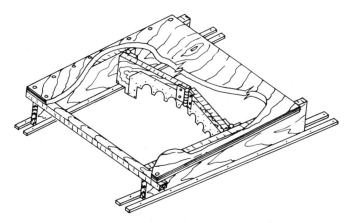

Fig. 109. Running, diminished: triple-hinge mould

ensure that the running mould runs smoothly over exactly the same course each time.

The method produces less distortion than the two previously described, what distortion there is being symmetrical about the centre line. The centre hinge should be let into the stock so that the profile is in line with the pivot point within the hinge, otherwise the stock will part as it closes, thus enlarging the centre feature of the run. To run on surfaces of changing curves, the stock is connected to the front of the slippers and the underneath of the slippers cut away so that only the back and front of each touch the running surface, as with a peg mould.

Diminishing a run in width and height

One-half of the section is run at a time, as in Fig. 110. A right angle is formed by boxing two rules

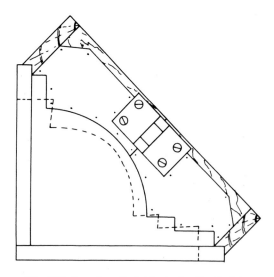

Fig. 110. Running, diminished: double diminish

cut to the diminish: one, half the width of the moulding; the other, the height. The bearing edges of the slippers are notched to fit on the rules. The distortions caused by this method are an angling of arrises and a flattening of the full section when

assembled. This will be more noticeable with some designs than others.

Figure 111 shows how the overhead rule to achieve the diminish in height is effective with certain sections, the best results being obtained in rounded ones. Flat sections tend to be flattened in the centre far more than at the sides. Fillets are similarly distorted: level ones slope out of level in towards the centre. This method is an adaptation of Fig. 109 with modified hinges to allow the double pivoting at each hinge. As the bottom of the stock tends to be lifted from the bench at the narrow end of the run, it is better to incorporate a moulding ground in the section.

RUNNING DOWN

The operation of running down lengths of moulding on a SPOT BOARD (Fig. 112).

RUNNING MOULD

See MOULD, RUNNING.

RUNNING IN SITU

Running in neat Class A Plaster

This should never be carried out *in situ*. The only times that a solid plasterer should run in neat plaster are when he is running down on the spot board. The expansion of a neat Class A plaster can cause lengths of moulding to part from the backing coat and eventually to fall. Also it would be extremely difficult, due to the excessive expansion, to obtain a good fin-

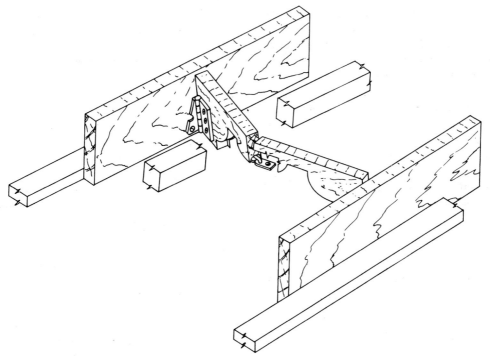

Fig. 111. Running, diminished: triple-hinge mould for double diminish

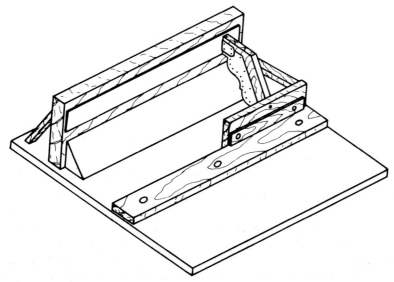

Fig. 112. Running down (solid): a length of cornice for returns

ish on mouldings in awkward positions. However, if one must do so, then a strong hard background is essential, a good muffle run core where possible, and plenty of keying with galvanised nails half driven in to the background to assist the bond.

Running in lime putty and plaster (Class A)

The gauge for this should never be weaker than 50% plaster, 50% lime putty. The two materials should be thoroughly mixed together before use, any unmixed portions of lime will remain soft until it dries out; then it will either flake off or crack and disintegrate.

To mix: estimate requirements and divide by 2. The first of the two parts will be lime putty, place this on the spot board and form into a large ring so that the putty will form a circular wall. Into this circle add the second of the two parts, clean water. Now add (Class A) plaster of Paris by holding it in two hands and sprinkling it on to the surface of the water. Continue till all the water has been soaked up by the plaster, then mix together using a hawk and laying trowel.

Whether one is running a cornice, a panel mould, or whatever, the method of running is basically the same. The plaster should be of a thick creamy consistency. It should be applied by trowel evenly over the area to be run, with slight emphasis on the front of the run and the uppermost member. Pass the running mould along the entire length – one man running, the other feeding – and continue till the moulding takes shape. Any surplus material should be used to core out other runs; one should always gauge too much rather than not enough. A second, slightly wetter gauge is next required. It must contain the same proportions of putty and plaster, and it is applied by trowel evenly to the shaped but unfinished moulding.

Continue for two or three more runs until the moulding reaches perfection as far as shape is concerned. To get a smoother finish, allow the moulding to harden, then with a clean running mould and water on the rule and screeds, give a good hard run over. This should result in an excellent finish.

The running mould must be thoroughly cleaned before each run.

Some craftsmen indulge in a practice known as *fisting* when running in putty and plaster. To do this they push the running mould backwards and forwards and at the same time work material into the member which requires it with their fingers. Fisting should take place once the moulding has been reasonably well formed, and the use of a barrier cream, grease, rubber or plastic glove is recommended. The action of running in both directions will greatly assist in the formation of a good shape but care must be taken to see that the metal profile is well fixed to the stock. It is not a universally accepted practice and many plasterers will not use this method at all.

Running with cement and sand

The core should be run out to a near-perfect shape using a muffle. As it is almost impossible to run with a stiff gauge, a thin mix is applied, run over with the running mould, and then stiffened by applying driers. Therefore two batches of material are gauged, one wet and the other dry at 1:2 or $2\frac{1}{2}$ cement: sand. A thin coat of the wet mix is applied to the core. This can be laid on or thrown on, but whatever method is used, care must be taken to see that the first coat is well worked into the keyed core. To ensure that the run is clear, the mould should be run along its entire length. Returning to the start, a coat of the dry mix is thrown on or spread on lightly. When the driers have become dark through absorbing moisture, the mould is run over the complete length. Both mould and rule are then cleaned.

A further application of the wet mix is made and the procedure repeated until a perfect shape has been obtained. This should be keyed and left for a period of 24 hours. The final mix – which is usually 1 part OPC: 2 to $2\frac{1}{2}$ parts sand – must not be stronger than the backing.

Ideally, the sand for finishing should be fine, washed on site then taken to a place where it can be left to dry out before use. The quantity for the finish run is estimated, and the complete mix turned over dry. When a uniform mix has been obtained, half of the material is mixed up to a wet slurry while the other half is left dry. The finishing is applied and run off in the same way as the core. When it is sufficiently firm small, shaped wooden floats are passed over the entire area to produce a sandstone texture.

Wherever possible mitres, breaks and returns should be filled out while the running operation is in progress and then finished freehand with the aid of wooden joint rules and small floats. (See also specific moulding to be run.)

Running in Keene's cement

Keene's cement is not currently commercially available but this entry may be of use in restoration work. This plaster has a slow continuous set. There will not, therefore, be any expansion to assist in obtaining a perfect finish.

A good run core is essential. This should be run out to a muffle and should consist of equal parts of Class A plaster and sand.

The types of moulding that may be formed in Keene's cement are the moulded head to a plaster skirting, angle beads, flush beads, solid pilasters and columns, panel and projecting dado mouldings. In rare cases, internal cornices may also be formed in Keene's.

The superfine grade of plaster is the best material for running. However, if this is not available, a little of the coarser grade may be sifted to produce sufficient for the final run. Gauge the Keene's in clean water to a creamy consistency and, working from right to left, apply evenly to the moulding area. Work well into the well-keyed core. Pass the running mould over the entire length and continue the

process to the whole of the running area.

Return to the first length and build out gradually to the required shape. Always use creamy plaster, as stiff Keene's will tear off, leaving holes and gaps. When the shape is perfect everywhere and has become firm, gauge the finishing coat a little thinner and apply this gently to the moulding. Some plasterers use a brush for this application. Continue running until a perfect finish has been obtained.

Mitres may be roughed out during the running or they may be left until the Keene's is hard. If they are left, all projecting particles must be cleaned out on the day of running as by the next day they will be very hard. When running most Keene's cement mouldings, it is usual to run up good setting members. These form an excellent hard arris and make good setting screeds.

In general, Keene's cement mouldings are found in public buildings and in places where they can offer resistance to hard knocks.

Running rules, screeds, rebates, running moulds as for running in plaster of Paris.

RUNNING ON A BENCH

All running on the bench is carried out with plaster of Paris. If hard plaster or, indeed, glass-fibre moulds are required they are cast from plaster models.

The plaster bench and the running rule should be thoroughly cleaned and greased before running starts. Any set plaster adhering to the rule or bench will cause the mould to jump and loose particles will mix in with the soft plaster and be dragged along the face of the moulding by the profile. It is not always necessary to make good the surface of a bench before running, so long as the nib or slippers run smoothly over it. Moulds with nib slippers will run perfectly over moderately pitted surfaces. The plaster moulding will need to be securely fixed to the bench during running. This is necessary, especi ally with larger mouldings, as the finish to the moulding is obtained by the profile scraping the hardening plaster; the resistance offered by the expanding moulding would be enough to tear it from the bench. The expansion alone is sufficient to release the moulding from the sealed and greased bench surface and it therefore needs to be anchored by nails projecting from the surface of the bench.

If the moulding being run is to be taken up, one nail only is driven into the bench at the beginning of the run outside the length marked on the running rule. In this instance the finished moulding can be cut to length and taken up. If holes and marks have been left in the bench surface, the resulting bumps on the back of the moulding will prevent it from lying flat and should be removed at this stage. Both solid plaster mouldings and reinforced plaster casts will bend and distort while they remain wet, and if they dry out in this position they will break rather than return to normal. A moulding that remains in position throughout its term of use (e.g. a reverse mould) should have nails every 0.250 m or so along its length to hold it down firmly during the removal of casts, etc. A long, narrow reverse mould containing no reinforcement will probably break.

Though nails are used to hold the reverse mould down, the bench must still be adequately greased to prevent it from being damaged when the mould is removed after use. The GREASE is applied evenly to the bench surface with a grease rag. Care should be taken to wipe grease into any deep holes that may be present. The running mould should be checked to ensure that it runs clear of the nails and that it is sound in itself, i.e. that it does not rock and that there are no loose members.

Gauging the plaster

In all but moulds of the smallest section, the maximum amount of running time must be gained so that the section of the moulding may be fully built up with the main gauge. This is not just for the sake of convenience: it is essential so that the finish may be achieved with a second, thinner gauge before the main gauge expands fully with the completion of its set. If this is not brought about, the main gauge may set fully while parts are being built up with the second gauge. When the profile hits these fully set and expanded parts the mould will jump, chatter or stop completely, depending on the size of the section.

In this situation it may even be necessary to use a third gauge to produce a finish on the places built up with the second. Even if a finish is achieved in spite of this, it will have a patchy, flaky surface that will become evident under the wear of many casts. This state may also be brought about by not gauging enough plaster to run with. It is essential to have just a little too much plaster rather than not enough.

The gauging container must be well rinsed, and only clean, crystal clear water used. To gauge plaster to a creamy consistency, sprinkle the plaster into the water at the rate at which it sinks. Carry on doing this until the plaster no longer sinks but shows itself just above the surface of the water over about half the area of the container. This will mean that the water is 'full' of soaked plaster, and the mixture – when stirred – will be within a minute adjustment of the required consistency.

Adjustments to the mix should never be made by the addition of water as this will accelerate the set. Plaster, however, may be sprinkled on to the surface while stirring. Pockets of dry, unsoaked plaster will cause lumps to form. Plaster of Paris is gauged with the hand. The plaster is sifted into the container through the fingers so that lumps may be rejected. The hands are greased while the plaster is soaking. Then, while stirring with the hands, any lumps may be detected and dealt with.

(A hallmark of the craftsman in solid plastering is a clean pair of hands. Materials such as Carlite, Sirapite, cement and lime are caustic in effect and might well make the plasterer a hospital case within the week if he were to use them as the fibrous plasterer uses plaster of Paris.)

With large sections it is advisable to get as much plaster on the bench as quickly as possible within the working limits of the plaster. The plaster applied to the bench must stay within the width of the section being run. That outside the section will become unworkable, cause mess and, inexpertly worked, will creep under the nib and slipper of the running mould to cause GATHERING ON. The volume of plaster available for the run will be reduced.

Plaster gauged correctly will give twenty minutes' running time. With small mouldings it is not necessary to place the plaster in the path of the running mould until it has picked up sufficiently to form peaks. It will then be stiff enough to stand up and be cut by the profile without delay. The fewer times the mould is run over, the better.

Incidentally, if the plaster is gauged to this degree of thickness in the container a satisfactory run will not be achieved. The plaster will set much more quickly, giving less running time. It will be denser and will behave differently during its set and, when cut by the template, will produce a coarse, woolly surface rather than a smooth, silky one. As the plaster picks up, the members contained in the section may be built up by repeatedly applying plaster and passing the running mould over. Each time this is done the running mould should be cleaned by removing the build-up of plaster from before and behind the template and from the nib and slippers where they make contact with the bench and running rule. In the latter stages of running larger moulds, water may be used as a lubricant on these bearing surfaces, care being taken not to splash any on the moulding.

The plaster on the bench will pick up due to its having been agitated during use. The plaster in the container, also picking up but to a lesser extent, may be knocked back, a piece at a time, by taking some out and beating it to a creamy consistency in a bowl.

It is then placed in any holes in the run, and at the beginning, to be carried by the running mould the full length of the run. In the early stages, surplus plaster carried to the end of the run and pushed off to either side may be replaced but, as the plaster picks up, this much-agitated plaster will be too stiff to use and must go to waste.

The section should be almost built up by the time the plaster is stiff enough to require knocking back. This knocked-back plaster becomes very workable, acting as a second gauge, and is applied to any holes and imperfections in the run. Once knocked back, however, it sets very rapidly. It is essential, therefore, to be completely ready to run the mould over before the plaster is knocked back. With correctly gauged plaster the span of the set should make it possible to knock back different portions of the plaster left in the container for about half a dozen runs between the time the plaster has stiffened sufficiently to need knocking back and when it is too stiff to be re-tempered. If the plaster is too stiff to roll in front of the template as it is carried along by the running mould, it will tear the surface and cause damage.

At this stage the finish should be good enough to be brought to perfection with a small new gauge of creamy plaster. The first application of this new gauge should be lightly rubbed into the surface of the hardening moulding with the fingers, care being taken not to penetrate the new plaster, as rubbing the moulding will knock it back and wear it away.

(If it happens that parts of the section are not properly built up at this stage, the second stage should just be allowed to soak. Thick, stodgy plaster is dredged from the bottom of the container and used just to fill the holes. This should allow the section to be built up in two or three runs. Too much plaster can be deliberately added to the gauge in order to provide sufficient thick plaster. When the section is complete the plaster may be stirred to a uniform consistency and used as a normal second gauge.)

At this stage a green suction will be provided by the first setting gauge of plaster. This will suck the water from the second gauge quite quickly so it is essential to sprinkle on the soft plaster and pass the running mould over it as soon as possible. To obtain the best possible finish, the soft, wet plaster should be scraped over the expanding moulding by the profile.

If the bench and bearing surfaces of the running mould have not been kept clean, the plaster that will have crept under the slippers and nib will have formed two screeds. These will be expanding and sucking water from the second gauge of plaster, thus thickening themselves and raising the running mould from the bench surface. This will cause the template to be lifted and a film of plaster will creep under it instead of being cut off by it. This effect is known as gathering on. However, to remove the screeds of plaster from the bench at this stage will cause the running mould to drop too drastically, giving the profile too much to cut off. This will result in chattering or in totally preventing the mould being run over the section.

Even when all these points have been mastered, it is not sufficient merely to have produced a good-looking finish. If the section is to be a reverse mould, from which many casts are to be taken, it will be subjected to great wear. The arrises, normally the last to be built up by the knocked-back – and thus weakened – plaster, need to be as strong as possible. To provide a good base for shellac, the finish in general needs to be strong and have been cut from good plaster – not from plaster that is slimy or cheesy.

Larger sections may require muffling and/or coring out. See MUFFLE; CHATTERING; CORE.

RUNNING UP COMPOUND CURVES

See ARCH.

RUNNING UP MITRES

See MITRE, RUNNING UP.

SADDLE

A fibrous plaster back which holds pieces of a reverse mould in position.

SAFETY AT WORK

See Appendix 1, p. 277.

SALT AS AN IMPURITY

See SAND.

SAND

Used as the main hard aggregate in plastering mixes. In external moulded work – cast or run *in situ* – mixed with cement. In external finishes and renderings, mixed with cement or cement and lime. In internal undercoats, mixed with cement or cement and lime or Class B gypsum plasters. In floor laying, mixed with cement.

Its function is that of an inert filler. It reduces proportionally the contraction and expansion of the binders, both initially on setting and hardening, and subsequently, on wetting or drying.

A strong, water-resistant external rendering tends to crack on certain backgrounds. It will shed water into the crack, resulting in water penetrating to the background and being trapped behind the rendering with consequent freezing-water and damp faults. Sand can be used to produce a more porous, absorbent and weaker surface, thus reducing cracking, and has the ability to soak up and hold rain without its penetrating too much to the background and to allow subsequent easy drying out.

Types for plastering

Sand may be taken from pits, rivers or seas, or produced by crushing coarser aggregates. It is almost impossible to remove salt from shore sand but it has been found that sand dredged from the sea bed and washed will be virtually salt-free.

Impurities

Clay

This will be present in sand from all land sources. It is important that a sand contains a small proportion of clay with the silt but if this is above 4 to 5% it can cause many faults:

1 The set of Portland cement is retarded.
2 The retarders in Class B plasters are attacked and the set therefore accelerated.
3 Cracking and blowing occur due to:
   (a) a greater amount of water being needed to lubricate the mix and a consequent increase in the amount of moisture movement on drying out;
   (b) shrinkage of the clay itself on drying out. This can cause crazing if it is distributed evenly throughout the mix and blowing if present in balls.

Loam (humus, organic matter)

Will delay or prevent the set of Portland cement.

Soluble salts (sea salt, gypsum, nitrates, etc.)

Can cause:

1 EFFLORESCENCE.
2 Corrosion of metal work.

3 Salt in a mix will, being hygroscopic, tend to attract moisture from the air, causing dampness of the plaster.
4 On rare occasions the salt may crystallise within the structure and so disrupt it.

Iron

Iron and iron pyrites may oxidise to cause staining and blowing.

Grading

A sand is passed through a series of sieves to determine the number of particles of various sizes. The proportion of the various sizes will determine its suitability for plastering and its particular use. Within the grade selected (e.g. floor laying, undercoats, finishing), a well-graded sand is one of varying rather than uniform particle sizes. A poorly graded sand is one overweighted with a particular particle size and is known as a uniform sand. Such a sand can cause many faults, for the lack of small particles to fill the voids between larger particles will necessitate the use of more binder. This adds expense and may produce too strong a mix which gives rise to cracking; water will run easily from the mix, taking the binder with it; any keys will not be sharply undercut and there will be a high percentage of droppings.

Bulking

This is a term given to a damp sand whose increase in volume has been caused by the particles sticking together, thus preventing complete compaction. Measuring a damp sand by volume without taking this into account will result in too little sand being introduced into the mix. To obtain the percentage to be allowed for, a known volume is shaken with an excess of water and, thus compacted, measured again.

Silt

Very small particles of sand, below No. 200 sieve size. A small proportion of fines – silt and clay – is beneficial to a sand. It will plasticise the mix, producing good workability and compaction with a minimum of binder. Too much will give rise to the faults listed above, under *clay*.

To determine the quantity of fines in a sample of sand, it is put into a glass jar with a surplus of salt water (5.00 ml salt to 0.50 l water) and shaken vigorously. Within ten minutes, the silt will have settled on top of the sand; the clay finally settles on the silt some hours later. The depth of sand, silt and clay can then be measured. Ten per cent of combined fines will make the sand unsuitable for use.

Sharp sand

Sharp, angular grains. The best type for plastering, producing the strongest mix.

Soft sand

Round, smooth sand, producing a weak mix.

SANDWICH MOULD/DOUBLE OR TWIN-STOCK MOULD

A metal profile sandwiched between two wood stocks. It is used when running vertical mouldings on either side of a panel or opening and eliminates the necessity for making two running moulds, one right-handed and the other left-handed.

The two stocks make it possible to run upwards on both sides, with the running rule on opposite sides of the moulding. With an ordinary running mould, one side would have to be run downwards (not good practice) or un up with the stock in front of the profile. As the plaster hardens, the metal profile can be forced backwards away from the stock unless they are well fixed together. This will cause distortion to the moulding, and plaster will get in between the profile and the stock – making it difficult to clean.

SCAFFOLDING

Falls from a height cause more deaths than anything else on a building site. If you can fall more than 2 m from a temporary structure – ladder, hop-up or whatever - then that structure will come under the law on scaffolding. All scaffolding work must be carried out under strict supervision by what the law calls 'a competent person' and scaffolds should only be erected by competent, trained scaffolders. An employer or the person in control of construction work must make sure that the scaffold is safe before workers are allowed to use it.

Scaffolds must be inspected by a 'competent person' at least every seven days. Any faults found must be put right. They must also be inspected after any addition, alteration or dismantling and after anything has happened to affect the strength or stability. A checklist includes: footings, standards, ledgers, bracing, putlogs and transoms (wrongly spaced?), couplings, bridles, ties, boarding (bad boards? trap boards?), guard rails and toe boards, ladders. A written report on the inspection must be prepared within 24 hours and kept on site until the work is completed. It should be kept in the office for three months.

Safe use of scaffolds
Fig. 112B indicates the safe layout of guard rails.

- Do not make any alteration to a scaffold to gain access to work.
- The only person to make changes must be a competent scaffolder.
- The platform of a general purpose scaffold should be at least four boards wide, the working platform of a tubular scaffold 3 boards or a minimum of 600 mm wide, on a trestle 2 boards

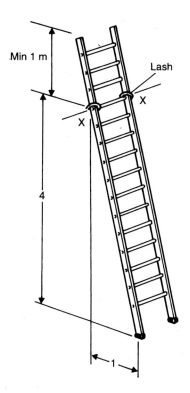

Fig. 112A. Ladder ratio, lashing points X

or a minimum of 430 mm wide. Never work from a platform that is not fully boarded.

Platforms, gangways, runs and stairs must be kept clear of unnecessary obstructions, materials or rubbish and from projecting nails. Slippery platforms, gangways, runs and stairs must be sanded or cleaned.

- Don't cause shocks by dropping loads.
- Make sure that all scaffolds, including 'independent' scaffolds, are securely tied.
- Avoid overloading the scaffold.
- Store materials for a short time only and so that the load is spread evenly.
- Look out for: traps (overhanging boards that will tip when trodden on), toe boards not in place, guard rails not in place, bad boards, boards with

too great a span, or components that someone may have tampered with.

Relevant legislation is: Health and Safety at Work Act 1974; Management of Health and Safety at Work Regulations 1992; Provision and Use of Work Equipment Regulations 1992; Construction (Health, Safety and Welfare) Regulations 1996.

Tower/mobile scaffolds (commercial)

Components of these systems are interlocking and include bracing and guard rails. The manufacturers' recommendations for erection and safety must be complied with. Wheels, on those that are equipped with them, must be locked when the scaffold is in position and in use. Outriggers can be used for greater stability. For internal use the height should not exceed 3.5 times the shortest base side, including outriggers. For external use the height should not exceed 3 times the shortest base side including outriggers. Always ascend tower scaffolds internally.

Scaffold boards

Scaffold boards should be straight-grained through the length and free from large knots. Fifty-mm thick boards must not be less than 200 mm wide but thicker boards may be as narrow as 150 mm.

Span: 32-mm boards maximum span 1 metre
38-mm boards maximum span 1.5 metres
50-mm boards maximum span 2.6 metres.

Ladders

A plasterer seldom works directly from a ladder but normally uses it only for access. Before you climb one, have a look at it.

- Is it in good condition? Ladders, if coated at all, should be varnished rather than painted so that any defects can be spotted easily. The wood grain should run with the members.

237

- Is it on a firm surface?
- Is it long enough and placed at angle of 75°?
- Is it firmly lashed?
- Does it extend about 1 m above the working platform?
- Do not climb on or off it if you cannot hold on to it.
- Never overreach.
- When using a stepladder do not use its top platform unless it has special handholds.

By law, trestle platforms and stepladders must not be used as a workplace above 2 m in height unless proper edge protection is provided. See Fig. 112A for the safe usage of a ladder.

Hop-up

This is a two-step-up device, the top step of which is a working platform at least 500 mm wide and 400 mm deep. You can buy ready-made hop-ups in aluminium or they can be hand-made from timber. They can be used with or without boards and provide access to the top of walls that are no more than 2.5 m high. They should be maintained in a sound and sturdy condition.

SCAFFOLDING

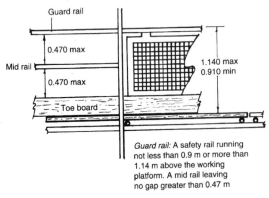

Guard rail: A safety rail running not less than 0.9 m or more than 1.14 m above the working platform. A mid rail leaving no gap greater than 0.47 m

Fig. 112B. Close boarded scaffold

Splithead

A splithead is a metal tripod with a slot to carry a beam member on the top of a centre pole of variable height. The bearers, which are timbers on edge, are wedged firmly into the slot with suitable pieces of wood to prevent any play or rocking. They are used to support a close-boarded working platform that can be adjusted to heights between 770 mm and 2 m. A minimum of four splitheads (one in each corner) is needed for a working platform and more will be needed for large areas. If the span is too great for the bearers, extra splitheads will be needed for intermediate support. If the span is too great for the boards, extra bearers and their splitheads will be needed for intermediate support for the boards.

- Place the working platform as near to the wall as possible so that you do not fall off.
- Are the pins sound?
- Make sure that the heads are not bent.
- Are the tripods firmly supported and free from rocking?

Trestles

Trestles are solely for use with scaffold boards and should never be used as ladders. Heights range between 2 m and 4.5 m. They are wide enough to take two scaffold boards or one lightweight staging. For strength, at least two tie rods should be fitted to either side. As with ladders, any coating should be of varnish rather than paint. Before using check:

- Are the trestles sound, with no missing or loose cross members?
- Are the hinges firm?
- Are the trestles fully open?
- Is the working platform less than 2 m high? (Otherwise toe board, guard rail and all scaffolding regulations come into force.)
- Is the platform fixed to the trestle?
- Is at least a third of the trestle above the working platform?

- Is there at least 450 mm clear space all round? and, when you have to move it:
- Dismantle it first.

Note: Trestles may not always be suitable for plasterers' work. Vigorous work in solid can produce a sideways thrust that the trestle is not sufficiently stable to cope with. Obviously, the higher the platform the greater the instability.

Mobile elevating work platforms

Operators of scissor lifts and cherry pickers must, by law, be trained in the safe operation of the machines. The certificate that is proof of this training should be carried by the operator when he/she is at work.

SCAGLIOLA

An imitation of marble, consisting of a hard plaster – traditionally the finest white Keene's cement – coloured with powdered dyes that are resistant to attack by alkali. The batches of coloured plaster are mixed to the correct shade and worked in such a way that the result is a true reproduction of any specific marble. The plaster is added to the coloured water and mixed to a thick, crumbly, dough-like consistency. This consistency ensures that the colours will not run.

The imitation is best made by working the coloured plaster in a way imitative of that in which the marble was formed, e.g. a laminated structure is produced by laying one flat layer upon another, repeatedly cutting in half through the layers, and placing one piece back upon another; the small lumps for the mottled fleck of granite by forcing the mix through a coarse sieve; veins by laying string and cotton soaked in dye across the face, the strings being removed after setting and the streaks subsequently filled.

In situ the mix is fashioned on a damped, well-keyed, strong sand and cement backing and forced well in. Although small casts may be solid and special blocks may be made for carving and turning, with conventional casts the coloured layer is the firstings, backed by conventionally reinforced seconds with, perhaps, extra or even metal reinforcements to combat weight in large casts. The effect being obtained may be observed by working a sample piece on glass and turning the whole over for inspection. As the coloured plaster is worked so stiff, many holes will be present in the face of the cast after rubbing down. Rubbing down produces clearer markings as slight merging will take place on the face during casting.

Filling is accomplished by repeated grouting and rubbing down with stones of various grades which need to be shaped when they are required to fit mouldings and enrichment. The cast is allowed to dry and the face soaked with linseed oil to fill the pores and so prevent the polishing wax soaking in. Poppy-seed oil may be used when the work contains white areas that will be yellowed by linseed oil. When dry, the face is polished – traditionally with paraffin and beeswax, although more up-to-date waxes are used; also even clear, two-pack polyurethane gloss varnish. This latter is applied thinly to the original dry cast and worked skilfully so that it does not give the appearance of an artificial coating over the face; this finish is the most suitable for table tops.

As the face cannot be pierced, all fixing is done from behind and the use of marble-fixing techniques, in fact, aid the illusion, as (at suitable intervals) do the types of joint used for marble.

See also CASTING IN KEENE's CEMENT.

SCALLOPED

A scallop-like feature found as NICHE heads, etc.

SCOTCH BRACKET

See BRACKETS.

SCOTIA

See MOULDING, SECTIONS.

SCOURING

The hard rubbing of the FINISHING coat, by cross-grain float and water in lime plaster. It re-tempers and compacts the surface. This will assist in obtaining a perfectly flat surface, also eliminate the possibility of cracking and crazing. See also FAULTS.

SCRAPED FINISH, EXTERNAL

See FINISH/RENDERING, EXTERNAL, cement plainface.

SCRATCH COAT

See RENDER COAT.

SCREED

A guide or ground, a backing on which a finish other than plaster or plasterer's material is to be applied, or a means of providing flat, smooth, bearing points for running moulds.

Floor and roof screeds

In roofing, a cement screed is normally covered by another material. Lightweight roof screeding is exfoliated vermiculite, substituted for sand and cement. These should receive a cement – sand topping after a period of 24 hours.

For the practical application of a floor screed see FLOOR LAYING AND ROOF SCREEDING.

Pressed screeds

A pressed screed in conjunction with DOTS is used as a guide in the formation of plain surfacing to curved backgrounds. Once the dots have become firm, a rule, which has been accurately cut to the curve, is held on the dots to ascertain the approximate thickness of the screed. The rule is taken away and the screed laid in position with the laying trowel. On completion, the rule is again offered to the dots, and the screed pressed into shape. All irregularities are either ruled off or built out until a perfect curved screed is formed (Figs. 64, 129, 130).

A rule or RIB used in the formation of a pressed screed can be a single rule, several rules fixed together or an accurately cast fibrous plaster template. Circular work is relatively simple as the line can be obtained from the centre. When elliptical, the major and the minor axis should be set out on the floor or bench, and the curve developed by one of the geometrical methods for forming an ellipse. If the template is made of fibrous plaster a lath on edge can be bent to the line and held in position by partly driven-home nails or brads. The lath and bench or floor are then greased and the fibrous plaster rib is cast to this. Should wooden rules be required, then either set out directly on to the timber or take a tracing of the entire curve needed and transfer it to the rule or rules and cut out accurately.

When working to grounds or pre-set margins an allowance may have to be made for the finishing coat, then a muffle may be cut for the rule, or a lath nailed on the edge of the curved rule. This will ensure that the floating is cut back behind the finish line. After the floating has been completed, remove the lath or muffle and finish to the final line. See also FLOATING TO CURVED BACKGROUNDS.

Screeds, for running in lime putty and plaster

These are used to provide a true surface for the bearing point, namely nib and slipper, of a solid running mould – a comparatively rare process today.

To obtain the position of a putty and plaster screed for running, hold the running mould at each end of the length to be run and mark nib and slipper points. Snap or strike long lines to these points so that there is a clear indication of the exact position in which the screed is required. The material used for running screeds is lime putty and Class A plaster; these two are mixed together in a 50:50 ratio.

Many plasterers prefer a stronger mix of approximately 2:1 plaster:putty. This material is then applied along all snapped lines to an approximate width of 50 to 75 mm, so that the line is central to the width of the screed. They should then be ruled out perfectly flat, using a good straight edge held at right angles to the screed and worked gently upwards with a side-to-side motion. Once the screed is seen to be in contact with the straight edge at all parts, it should be left to harden a little before being closed in by a crossgrain float.

Here, one should take a *little* putty and plaster on to one handhawk, then take off just a little with the toe of the float and apply to the screed, keeping the float flat to the screed all the time and working in long sweeps from side to side. This should produce a hard float-finish appearance to the screed. At no time should water be applied to a screed to assist with this, as water may soften the screed. Neither should the screed be touched at this stage with a laying trowel as this could cause ripples in the screed face.

For alternative methods see REBATE.

Wall and ceiling screeds

Narrow bands of plaster, usually floating material, applied to walls and ceilings to assist in the formation of a true and even surface. Generally they are set between a pair of dots that have been plumbed down. Once these dots have become sufficiently firm to withstand pressure from the floating rule, a small quantity of backing material is applied by trowel, held in a vertical position and with a pinching action, to the area between the dots. As soon as this material reaches the same thickness as the dots, it is ruled in flat to the dots.

To do this, hold the rule on both dots and at right angles to them, then work the rule gently up and down along the dot face. Having cut off all the surplus material, inspect the screed and, if there are places where the rule has not touched, fill them in again and repeat the operation with the rule. As soon as every part of the screed is seen to be in contact with the floating rule, and flush to both dots, trim the edges straight and leave until firm enough to use as a guide for the floating of the wall or ceiling area.

Basically there can only be two types of screeds, horizontal and vertical. With one exception, the preceding description would also be suitable for horizontal screeds ruled to vertical dots – the material would be applied with the same pinching action, but horizontally.

The screeds may occasionally be applied without the aid of dots. In this case they are still the same narrow bands of plaster, applied horizontally, vertically or both, ruled off with a floating rule till flat, and possibly plumbed. When they are adjacent to a door or window frame, they should be margined or ruled in to it.

The position of wall screeds will once again depend upon the approach of the craftsman. One will prefer horizontal, another prefer vertical. Some plasterers work to four screeds on normal-sized walling. The two vertical screeds are applied first, then two horizontal ones – one top, the other bottom – are ruled off to the former and finally the centre gap is floated and ruled off to the horizontal screeds.

This system may be termed box screeding or picture-frame screeding.

Types of wall screeding

See FLOATING.

SCREED BEAD

See BEAD, METAL.

SCRIBE

See MITRE.

SCRIM

82 mm or 100 mm wide jute canvas, usually in rolls of approximately 80 m. Is used to reinforce joints in plasterboard ceilings and partitions. Metal scrim or strip mesh, normal or self-stapling, is used to reinforce points of potential weakness and joints in wood-wool slabwork.

SEASONING

The treatment of a surface (model or mould) from which a cast is to be taken. See under the material of the mould: COLD-POUR COMPOUND; GELATINE; GLASS FIBRE; PLASTER; POLYVINYL CHLORIDE; WAX, MOULDING. For clay squeezes see REPAIR AND RESTORATION.

SECONDS

See CASTING IN FIBROUS PLASTER, general method.

SEPARATE FLOORS

See FLOOR LAYING AND ROOF SCREEDING.

SETTING

The stiffening and hardening of any Portland cement or gypsum plaster gauged material.

SETTING COAT

The internal FINISHING coat.

SGRAFFITO

See FINISH/RENDERING, EXTERNAL.

SHELLAC

Shellac is made from lac, a resinous secretion produced by the lac insect on the twigs of various Far Eastern trees. To make shellac, the lac is processed and formed into thin, flat sheets which are broken up into flakes, sometimes referred to by plasterers as 'crystals'.

It is used in the seasoning of porous surfaces from which plaster, cement and sometimes resin casts are to be made. It is applied in liquid form with a brush to provide a sealing coat. The brushes vary from 12 to 100 mm, depending on the fineness of detail in the item to be treated. It may be bought in liquid form as shellac, knotting or button polish. If shellac 'crystals' are used the flakes are put into an earthenware or glazed container and methylated spirits poured over them until they are just covered. They are left to soak, being stirred from time to time, until dissolved – a process that usually takes about three days.

In plastering, all forms of shellac are thinned with methylated spirits and are therefore extremely quick-drying. This necessitates shellac brushes being kept permanently in the pot to avoid their hardening. When possible, the shellac pot should be covered to keep out dust and foreign particles. Ferrous metal left in contact with shellac will turn its light honey-brown colour very dark and eventually black. Although this has no real detrimental effect, the lighter colour is preferred, and trouble is taken to provide string-bound brushes and non-metal pots.

Application to plaster

There is no need for the plaster to be dry – shellacking can take place as soon as it is fully set. First the plaster surface should be brushed off to remove any dirt and foreign particles. To effect a seal the surface should be shellacked until it shines. The shellac should be tested on a comparable piece of plaster and thinned with methylated spirits so that a uniform shine will be achieved in three coats, the first of which is a thin primer and the second and third thickened by the addition of fresh shellac.

The rapid evaporation of the meths makes testing necessary, as the shellac can be too thick for use within only a half-hour of the previous thinning. The shellac should be applied thinly and efficiently with even brush strokes, care being taken to allow no tears or runs to form on vertical edges and internal

angles or other detail to become filled. When each coat is dry – only a matter of minutes – any 'bits' that were in the shellac or have subsequently settled on the wet surface should be removed with a dusting brush or by rubbing with the hand or a piece of canvas.

Any coating applied to a plaster surface containing sharp, cleanly cut members will dull it to some extent. It is not, therefore, desirable to paint on thick, treacly, obliterating layers; these will even change the texture of plain surfaces. When a surface contains more water than would be normal in a setting plaster it may be too wet to be shellacked. Shellac applied will congeal but not adhere, and will be rolled off by the brush strokes that are intended to apply it. Overwet surfaces may have excess water drawn from them by sprinkling on dry plaster and subsequently scraping it off.

Application to timber

The drying period of shellac on wood is comparatively long – anything up to one hour per coat. If several coats are to be applied it is common practice to apply a coat evenly and set light to it. This burns off the meths and causes the shellac to set immediately. This procedure saves time but tends to produce bubbles and wrinkles in the surface – the greater the thickness of shellac the worse this effect. Great care should be taken, however, as meths burns with a clear flame and the unguarded application of a subsequent coat can lead to the brush catching fire. The whole procedure is hazardous and not to be recommended.

Burnt shellac

Used as an adhesive, chiefly to stick enrichments in their beds when dressing models. It is produced by further stiffening thick shellac by burning off some of its meths content. This is done by placing the shellac in a small tin and setting light to it. The flame is smothered by placing a flat cover over the tin once it is judged that the shellac has reached the correct consistency, allowing for it to thicken further on cooling. When sticking to plaster surfaces, it is advisable to prime the keyed surfaces first by applying one medium-to-thin coat of shellac.

White shellac

Strongly alkaline plasters, such as non-expansion and special hard casting plasters, react with brown shellac to turn it bright purple. This is not detrimental but is considered undesirable as a general rule and, with this in mind, white or clear shellac may be substituted.

SHELLING

1 Caving in of firstings on a cast. See CASTING IN FIBROUS PLASTER, general method.
2 Finishing coat flaking from the floating in solid work. See FINISHING, summary.

SHINGLE

See AGGREGATE.

SHOE

1 A metal plate on a GIGSTICK. See SPINNING.
2 A bearer on a solid running mould. See RUNNING IN SITU.

SILT

See SAND.

SISAL

A natural fibrous material made from the sisal grass plant or American aloe. Used in plaster mixes as reinforcement. See also HAIR.

SIZE

A solution of glue size added to gauging water to retard the set of plaster of Paris. It inhibits crystal growth in plaster; a bonus is that the slower the set the harder the plaster will be. It may be made in GELATINE or from a proprietary glue size which will contain an anti-fungus and mould-resistant additive. A solution of the required strength will gel in the container but this may be prevented by the addition of a small amount of lime. The size container is provided with a measuring pot; the solution should be stirred thoroughly before each use.

Use

Chiefly to retard the set of plaster for casting. Size is not normally added to plaster used for running. It will cause the plaster to become hard and brittle during the last few runs when it is necessary for the main mix to remain as rubbery and cheesy as possible. The plaster cannot be knocked back efficiently and will often contain set granules that will be torn along the run by the profile. However, none of this will matter when running up a core with a muffle.

Size can be used to retard a backing coat of plaster of Paris gauged with sand or lightweight aggregate.

See also SODIUM CITRATE.

SKIMMING

The setting or FINISHING coat.

SKIN MOULD

See under MOULD, REVERSE.

SKIRTING

Running in plaster

Only the moulded members of the normal skirting should be run. The plain section or plinth is applied after the completion of the running and finishes to a setting member run up in position with the moulded section.

The first operation is to cut a plinth gauge. This is a piece of straight timber, 1 m long by 75 mm wide and 25 mm thick. At one end a piece is cut out to a size of 3 mm less than the height and thickness of the skirting. The walls of the room can be set before the skirting is finished, but it must be ruled in straight horizontally where the nib of the skirting mould will run. The backing for the skirting should be strong gauged plaster and sand or Keene's and sand. Several coats are applied until it is in contact with the plinth gauge. It should also be ruled in straight horizontally.

The rules are now fixed in position on the plinth. The running mould is offered in position at both ends of each wall. Mark where the slipper rests on the plinth, and check the height from the floor. Lines are snapped on these points.

A rabbet is fixed to the underside of the slipper, ensuring that the thickness of the lowest member is sufficiently back so as to form a setting member for the plinth (Fig. 113). The running off should be as for lime putty and plaster or Keene's cement, whichever is specified. Keene's is the better material because of its hardness.

When the running has been completed, all rules are removed. All material that has accrued above and below the finished mould is cleared and all mitre completed. The plinth is set to the setting member.

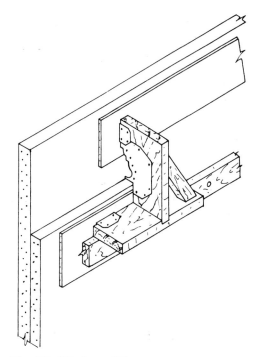

Fig. 113. Skirting (solid): running the moulded top

To check that this is square, hold a short rule vertically on the finished plinth and check the margin between the wall and the front edge of the rule. This should be parallel at all points above the finished skirting. The finish of the plinth will normally be the same material that was used to run the moulded section.

Coved skirting

Formed in either granolithic or Portland cement and sand. They are usually run on grounds, which may be either battens or screeds, by the use of skirting or a coving trowel. When using the former, an upstand is run as an extension of the cove. The bottom edge of the wall batten may be chamfered so as to produce a splayed or weathered top member to the skirting (Fig. 114).

Internal vertical angles may be coved in a similar

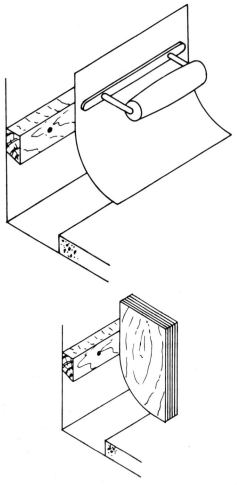

Fig. 114. Skirting, coved (solid): coving trowel and timber gauge

way, using a coving trowel to produce a flush cove between walls. The material used in this case should be the same as used to finish the wall surfaces.

SLAKED LIME

See LIME.

SLAKING

See LIME.

SLIP-PIECE

See under MOULD, REVERSE.

SLIPPER

See MOULD, RUNNING.

SODIUM CITRATE

A solution added to gauging water to retard the set of casting plasters. See also SIZE.

SOFFIT

The undersurface of features such as BEAMS; ARCHES; balconies, stairs, etc.

SOFT SOAP

Used both as a release agent in waste moulding and for general washing of wax moulds and the like.

SOLID PLASTERING

The application of a wet plaster mix *in situ*.

SPANDREL

A three-sided area, the sides of which may be curved, formed between features, e.g. beneath the string of a staircase, at either side of the curve of an arch, at the corners of an Adam ceiling, etc.

SPAR

White and light-coloured stone chips.

SPAR DASHING

See FINISH/RENDERING, EXTERNAL.

SPARROW PECKING

A KEY for solid plastering formed on a smooth, dense background with a special bit in a mechanical hammer.

SPATTERDASH

A mix of Portland cement and sharp sand, usually 1:2. Prepared as a thick slurry and thrown on to dense backgrounds to provide a key coat. Also provides even suction. The impact of throwing gives better adhesion than does spreading.

SPINNING

On flat surfaces – benchwork

This is the term applied to running circular work where the running mould is fixed to a pivot so that the mould revolves round it. As can be seen from Fig. 115, the set-up can range from tight circles where the stock of the running mould is fixed to the pivot and turns about itself, as with (a) Dishes (see also domes, Fig. 38), through an extended stock for small circles, as (b), to the use of an arm, termed a 'gigstick', fixed in line with the face of the stock in order to extend it to the pivot (c). (See also Fig. 116.)

It can also be seen that many arrangements exist for providing a pivot. In (a$_1$), (b) and (c) a metal plate fixed to a soft surface prevents the forces acting upon the pin during the spinning from causing it to work loose. (a$_2$) shows a pivot block, comprising a pin in a block of wood that is wadded to the running surface. With large sections that will impose great forces on the pin and cause it to work loose in end-grain timber, it is advisable to pass a substantial pin through a metal plate fixed to the block and also use a substantial shoe on the gigstick to fit over the pin, such as half a hinge (Fig. 116).

Points in horsing

1 The gigstick must be substantial enough to resist bending in all directions – even to the extent of constructing a T, angle, or box section and extending the slipper in order to angle a suitable brace from it to the gigstick (Fig. 116). Sagging of the gigstick and, therefore, the pulling in of the running mould towards the pivot can be avoided on long-radius work by supporting it by one or more rails.
2 The pin must be in line with the mould's metal profile, otherwise the section of the moulding being spun will be narrowed.
3 Extra clearance between the stock and the profile will be required on the inside of tight curves to prevent the stock fouling the moulding being spun.

Operation

Running up the section requires all the relevant

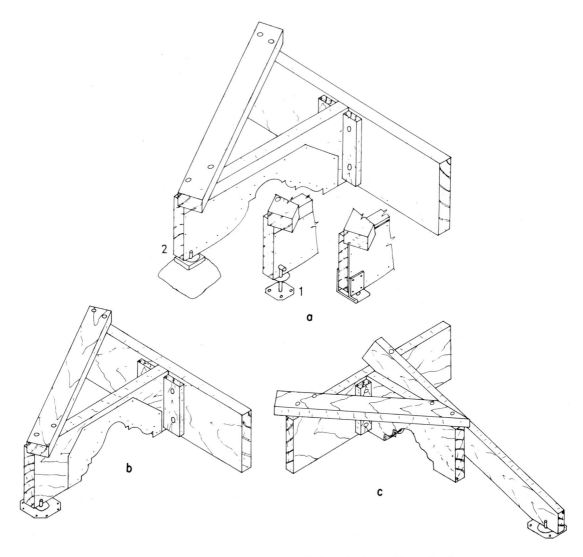

Fig. 115. Spinning: running moulds for spinning on flat surfaces

techniques of RUNNING ON A BENCH plus the following points:

1 If the section is to be removed it must be anchored to the bench by at least two vertical, headless nails off which the section may be lifted when it has set. (The practice of claying headed nails is unsatisfactory, as any force that will cause the moulding to move will force the nail through the clay.)

2 The section of a full circle is more easily built up in the early stages than a comparable straight one because the running mould, in revolving,

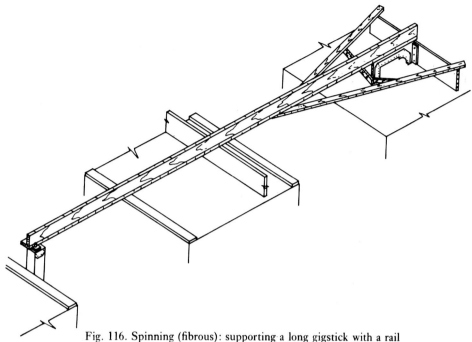

Fig. 116. Spinning (fibrous): supporting a long gigstick with a rail

will pass over the section in a ratio of somewhere in the region of 10:1 with each application of plaster; this effect 'feeds' the section. Similarly the second, finishing gauge can be used more effectively but, once the plaster starts to expand (unlike the straight run that will relieve itself mainly along its length), it will grow outwards away from the pivot, thus causing the outer facing surfaces to be cut but gathering-on to take place on the inner facing surfaces. This latter necessitates extra skill to produce a finish. The agitation of the plaster feeding the section, coupled with the suction afforded by the plaster in the section, will cause the plaster to stiffen and stick to the profile. When the plaster stops rolling, the running mould may be lifted clear of the section.

On curved surfaces – benchwork

Circle spun on a reverse mould for a hemispherical dome

(When possible, the moulding can be included in the section when spinning the dome.) A hemispherical DOME provides the simplest type of curved background on which to spin a circle because the concave surface of a dome is constant in all directions; therefore, a running mould constructed to fit it will do so at every point traced by the gigstick. The running mould may be one with a concave slipper and stock, or with a straight-bottomed slipper with a peg at each end, or a three-legged half-slipper running mould. Although a conventional pivot, shoe and gigstick will normally serve (Fig. 117), a knuckle join in the gigstick will have to be made if it fouls the curvature of the dome.

Circle spun on a drum with a changing curve (barrel vaulting)

A circle scribed from a centre point will show fore-shortening on the lower part (Fig. 118(b)), producing an ellipse when viewed from underneath, but, as the curve of the drum covered by the circle is usually very slight, the distortion is usually acceptable. (For those cases where it is not, see RUNNING CIRCULAR CURVES, on curved backgrounds.)

Also evident from Fig. 118 is the reason why the

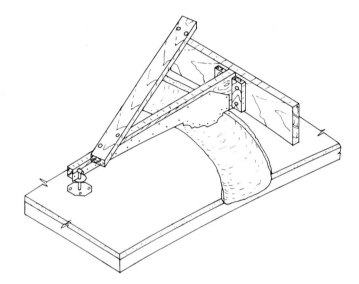

Fig. 117. Spinning: on a dome

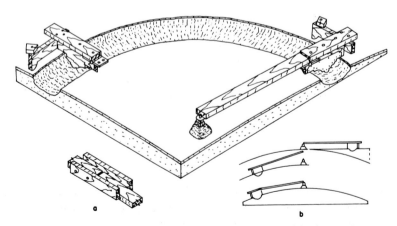

Fig. 118. Spinning: with a hinged gigstick on a curved background

gigstick needs to be hinged in order to allow its end to remain in contact with the pivot block while the running mould remains in contact with the surface of the drum. The surface traversed by the circle changes four times: it will have a curved radius with a straight tangent at the lower point and a straight radius with a curved tangent at the high point. A three-legged half-slipper running mould is required because a straight slipper, though fitting the drum at the low point, will rock at the high point. A concave or pegged full slipper to fit the curve at the high point would hold the stock off the surface at the low point. The slipper is cut away so that only its end bears on the running surface, and the bottom of the stock is feathered so that the profile remains in contact with the drum.

No practical method exists to keep the profile perpendicular to a changing surface. A profile horsed to suit the drum at the lowest point will lean back slightly out of plumb at the top, so reducing uniformly the height of the moulding. However, this distortion is slight enough to be acceptable and may be reduced by half by leaning the stock forward out of the pependicular by half the amount it will lay back at the top of the drum. The slight twisting of the gigstick may be taken up in the shoe at the pivot. When help is not available to hold the end of the gigstick down on to the pivot, the head may be left on the nail or screw used as the pin. (It is easier if the gigstick is hinged before it is fixed to the running mould.) When a large section is to be run, the expansion of which may reveal transverse play in the hinge, splints may be arranged to allow the required hinging movement only. (See Fig. 118(a).)

Circle spun on a drum with a changing curve (elliptical domes, circular domes with semi-elliptical section)

The arrangement required is the same as in barrel vaulting but, as every part of the surface covered by the circle will be a changing curve, the stock is leaned forwards out of perpendicular to the drum in order to balance the movement of the stock before and behind the perpendicular.

See also RUNNING CIRCULAR CURVES, on curved backgrounds.

SPLINE

Ruling in a curved surface by bending a pliable rule over undulating templates or ribs. See also DRUM, double-curvature.

SPOT BOARD

A flat table of approximately $1\,m^2$, used to hold the plasterer's wet materials. It may be made of water-resistant plywood, blockboard or floor boarding held together by two cross-pieces.

SPRAY PLASTERING

See MECHANICAL PLASTERING.

SPRINGING LINE

The line from which an ARCH or ARCHIVOLT springs. The centre or centres are not necessarily positioned on this.

SQUEEZE

A mould, any impression taken in clay or similar material, or a plaster mould taken *in situ*. See also REPAIR AND RESTORATION.

STABILISING SOLUTION

A liquid for the treatment of friable surfaces: It is painted on liberally, soaks in and hardens. For external use resin finishes may be applied to the stabilised surface, but internal surfaces are treated with PVA before plastering.

STAFF BEAD

See MOULDING, SECTIONS.

STAIRCASE WORK IN GRANITE AND CEMENT

The application should be similar to the laying of granite floors or cement screeds in floor laying or roof screeds except that in most cases a form of shuttering will be required.

Terms used

Tread – the upper surface of a step.
Riser – the vertical member between two treads.
Nosing – the exposed edge of the tread of a staircase.
Going – the distance between the first riser at the foot of the staircase and the last riser at the top of the staircase, measured horizontally.
Going – of a step – the actual measurement between two risers, or length to tread.
Flight – a series of steps without a landing.
String – the member which receives the ends of the step or, in plastering, more often the exposed outer edge of the concrete steps which may have to be plastered. An enclosed string is where the steps butt into a wall and an open string is where the step ends stand exposed.

Pitch of stairs – in practice it is usual to consider the pitch of stairs to be a measurement that consists of two risers and one tread. This should be in the region of 575 mm, e.g. two risers totalling 350 mm and a tread of 225 mm. For private stairs the maximum rise is 220 mm and minimum going 220 mm.
Storey rod – a rod with a number of saw cuts on it, each cut representing the height of the risers. Central to each mark a nail should be half driven in to the rod.
Riser board – the part of the shuttering that will assist in the formation of the riser and give a guide to the levelling in of the tread; it should be bevelled at the bottom edge.
Stringer or string board – this is the board into which the riser boards are fixed. The setting out of a staircase is done by dividing the total going travel of the flight into an equal number of treads to the given size. Then the total height into an equal number of risers, also of the given height. There will of course be one more riser than there are treads.

Staircase work, closed strings

The position of the nosing of the top and bottom step are found by using the storey rod with a long parallel rule and spirit level. (See Fig. 119.) These two positions are clearly marked and a chalk line is snapped on both walls. The position of the remainder of the riser boards is checked by levelling in from the storey rod.

Each time, the height of the riser and width of the tread must be checked. Any variation may cause an accident when the steps are in use. The riser boards, with the bevelled edge facing downwards and outwards, are then wedged in position and should allow sufficient thickness for the finish of the tread and the riser.

Before commencing to lay the steps, the concrete base should be swept thoroughly, soaked and finally grouted; this includes the risers. The mix for granolithic toppings or finishes must be either 2:5 cement: granite or 2:1:4, cement: clean sharp sand: granite. For cement-screeded steps, the mix should be 1:3, cement: sharp sand.

The method of laying is the same as for floor laying and screeding, with just one exception. The riser is actually cast by ramming material down in the gap between the riser board and the concrete riser. It must be filled completely and the following day, when the form work is struck, any irregularities made good. The whole step is then rubbed down with a carborundum stone and the nosing slightly rounded. The mix for the making good can vary from $1:1$ cement : sand to $1:\frac{1}{2}:\frac{1}{2}$ cement : sand : granite dust.

The treads of granolithic steps are usually made non-slip by either the application of carborundum chippings when green, or by rolling or grooving again while still wet. The stud roller is so constructed that it will form small indentations in the tread surface and the tool for grooving may be something similar to a pointer or, more usually, will have several blades. Alternatively, inserts are made using other materials when the granite has hardened. For this work, rules are fixed to the concrete tread so that an area to the size of the insert is left clear.

Where balusters are to be fixed on completion, plugs may be fitted into the concrete so that the plasterer can work round them. When they are removed a hole will be left to take the foot of the baluster.

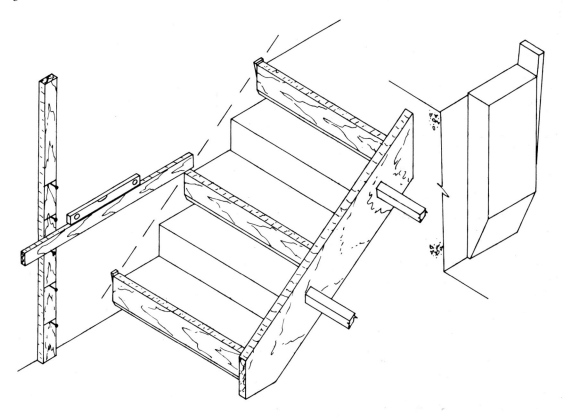

Fig. 119. Staircase work (solid): shuttering to steps

The treads must be protected when the finishing work is carried out on the following day. This is normally done by the operative using clean boards to walk and kneel on. Another method is to lay alternate steps. The bevelled edge at the base of the riser board is so as to allow freedom of movement for the trowel while the tread is being laid. Again any gaps left because of difficulty of application must be made good the following day; where a moulded or quarter-round nosing is required a reverse section of this is formed at the top of the riser board.

Staircase work, open strings

In this case, there is no brick or concrete wall to set out on. Therefore, string boards are used. The setting out can be done on these before they are fixed or it may be done *in situ*. The riser boards are nailed, screwed or housed and wedged into the string board and the fixing of these will be as for closed strings, fixed to a nosing line and levelled in to a storey rod (Fig. 119).

Usually part of the string will be finished in a matching band to the material used on the steps. To permit this to be done accurately, a long straight rule, the thickness of which must coincide with the material thickness, is fixed in position on the inside face of the string board. It will be parallel to the struck line and to the required measurement below the internal angle of the risers and treads. Material is pushed down into the gap so formed in the same way as for the risers. It is then made good and rubbed down the following day.

Staircase work, landings

Apart from the levelling in from the line for the top riser, the laying of either a cement screed or granite topping to a landing should differ little from a normal floor. See FLOOR LAYING AND ROOF SCREEDING; SKIRTING, COVED.

Nosing, moulded, to granolithic steps

The moulded section for the nosing is seated along the top edge of the riser board so that the top

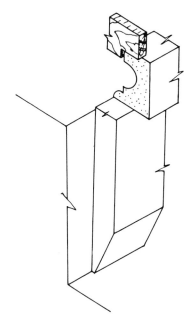

Fig. 120. Staircase work (solid): shuttering – riser board with moulded nosing

member is in line with the tread (Fig. 120). Alternatively, to avoid a feathered edge to the reverse mould, a small vertical member may be added over and above the moulded nosing, in which case the tread will be ruled in using a shoulder or rebated gauge. The reverse mould for the nosing may be made from hard pre-cast plaster or wood. When casting to this it should be oiled with paraffin and the granolithic mix pummelled in hard so as to eliminate the possibility of voids in the face of the nosing. The final finish to this member will be as for normal granité steps.

STILES

Cross-members to decorative arch or beam SOFFITS. They may be positioned one at either end of the soffit or they may break the soffit up into regular panels.

STITCHING

Joining a fibrous plaster cast to one being cast by placing it in a mould and lapping the new cast over its back.

STOCK

See MOULD, RUNNING.

STONE, CRUSHED

See AGGREGATE.

STONE, POLISHING

Carborundum: Used to polish granolithic steps when the shuttering has been struck. See also separate entry on CARBORUNDUM.
Snake: Used to polish imitation marbles.

STONEWORK, ARTIFICIAL

Any plasterwork that imitates a stone material.

STONEWORK, RECONSTRUCTED

Cast cement features made to look like natural stone by using the required crushed stone as an aggregate with white or coloured cement.

STOP, BIRD'S BEAK

The moulded feature, formed with a metal template or gauge, to stop a run or formed external angle and bring it to a sharp or pencil-round arris. They usually occur just below the cornice or architrave and just above the skirting. See ANGLE, Fig. 1(f).

STOP, SLIDING

These stops are placed on reverse moulds to form the ends of a cast.

Square stop

This forms a square end and permits casts shorter than the reverse mould to be made (Fig. 121). No production is lost if the stops are cut from the thickened-up end of the first cast. They are shellacked and then greased before each cast is made.

Mitre stop

This stop will form an end that is cut for a mitre, as in Fig. 122. Whereas little time may be saved in

255

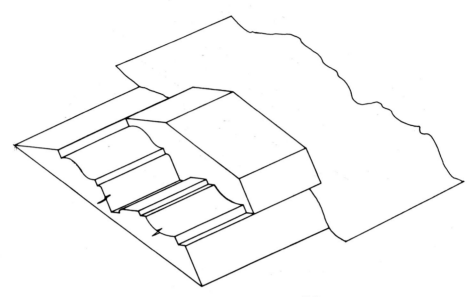

Fig. 121. Stop (fibrous): square or sliding stop

casting small sections pre-cut in such a way, considerable savings may be made when large, intricate sections are cast to mitre stops that have been accurately cut. The stop is formed by a normal mitre cut, as if the moulding were being cut for mitring.

The mitre cut on the stop will, of course, produce its negative on the cast, e.g. a stop cut as a right-hand internal mitre will form a left-hand external mitre.

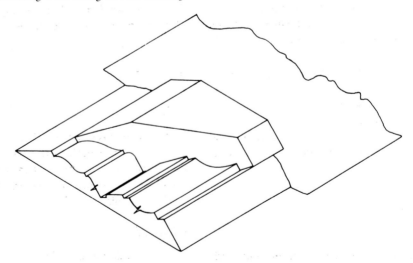

Fig. 122. Stop (fibrous): mitre stop

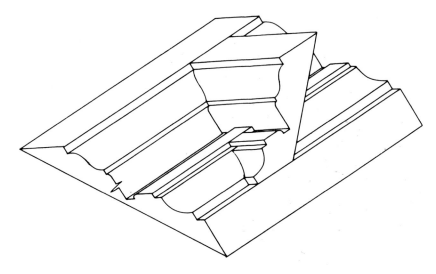

Fig. 123. Stop (fibrous): scribed stop

Scribed stop

This stop forms a return with an external mitre (Fig. 123). It is produced by scribing a piece of reverse mould into the main body of the mould. To do this, the main mould is shellacked and greased; the scribe is cut roughly to shape and bedded in soft plaster on the mould, care being taken to ensure that it is at the correct angle and that all members pick up. The internal mitre is then completed as the plaster sets.

When set, the stop is cut to length, shellacked and greased for use. This form of stop is used not only to form returned ends on mouldings but also in the body of features such as the breaks and returns in cappings. Because its width can be varied, it is used in preference to a cast piece mould for OVERDOORS, PILASTERS, etc.

STOPPING

Fibrous plaster and benchwork – filling and finishing MITRES, joints and FIXINGS; damage repair, etc.,

to new work. See under specific item to be stopped. See also MAKING GOOD; PLAINFACE, jointing.

STRAPWORK

Strap-like panel moulding making ceiling ornament; originated in Tudor times.

STRIKE-OFF

Of a mould

The edges of a reverse mould, off which the edges of a cast are formed, the perimeter strike-off and all other strike-offs contained by the case being ruled off them to the correct angle and thickness. See under particular mould.

Of a cast

The strike-offs are the strips of plaster on the back of a cast that are to come in contact with the fixings.

257

They are formed round the perimeter and across the cast, on top of the wood lath reinforcement and are brought to a correct, even thickness by ruling them in off the strike-offs on the reverse mould. See under particular cast.

STRING

See STAIRCASE WORK IN GRANITE AND CEMENT.

STRIP MESH

See SCRIM.

STUC

Gypsum plaster and sand used internally with an imitation masonry finish. Colours may be added and the finish may be smooth, float or drag. The joints may be pointed.

STUCCO

Most external finishes can be classified under this heading. See FINISH/RENDERING, EXTERNAL.

STUD ROLLER

See STAIRCASE WORK IN GRANITE AND CEMENT.

STYLOBATE

A continuous base on which a series of columns is placed.

SUCTION

The ability of a surface to absorb water from a plaster mix applied to it.

Control

Suction may be reduced, made uniform, or eliminated by applying a sealer (PVA), incorporating a waterproofer in a cement-bound undercoat, or by soaking with water. See FINISH/RENDERING, EXTERNAL.

Green suction

The early suction of a plastering mix which develops during its set or hardening.

SULPHUR MOULD

See MOULD, SULPHUR.

SYNTHAPRUFE RIW 232

See BACKGROUND.

TALLOW

See GREASE, PLASTERER'S.

TAMPING

See FLOOR LAYING AND ROOF SCREEDING.

TANK, LATH

See LATH TANK.

TEMPLATE

For the use of templates in plastering see ANGLE, EXTERNAL, FORMED; PROFILE, METAL; RIB.

TERRAZZO

A specialised finish to flooring, steps, dados, etc., consisting of coloured marble chips and white cement and polished to a high degree – today with an electric polishing machine. Pre-cast slabs can be obtained in a similar material.

See also FLOORING, TERRAZO.

THERMALITE

See BACKGROUND, building blocks – lightweight.

THROAT

The sunken bed or horizontal member of the DRIP section to a CORNICE. See also MOULDING, SECTIONS.

THUMB MOULD

See BRACKETS.

TILE, CLAY

See BACKGROUND.

TILE, GLAZED

See BACKGROUND.

TILE PARTITION, CLAY

See BACKGROUND.

TOOLS

The tools used in plastering are listed below. Where these are illustrated (Plates 1–4) this is indicated in brackets.

Applicator (4b)
Bob – centre (li), plumb (1h)
Bowl – gauging and stopping
Bradawl/pricker
Brush – dusting (1d), splash (1c), stippling/external texture (1e), two-knot (1b), tool, water, wire
Busk (3j)
Calipers – internal and external (3b)
Chisels
Cockscomb/french curve (3i)
Compasses, metal – adjustable dividers
Countersink
Darby (4e) – usually 1.2 m long
Dividers/scribes
Drag (3c)
Drills – for plaster, wood, metal, and screwdriver bits
Feather edge (4f)
File – for metal
Float – crossgrain (1q), devil (1p with a nail in each top corner), flute (1j, n), margin (1l), sharpened (1m, o), skimming (1p)
Gauges
Gimlet
Hair beater – an obsolete appliance used to separate and cleanse hair for haired lime plaster

259

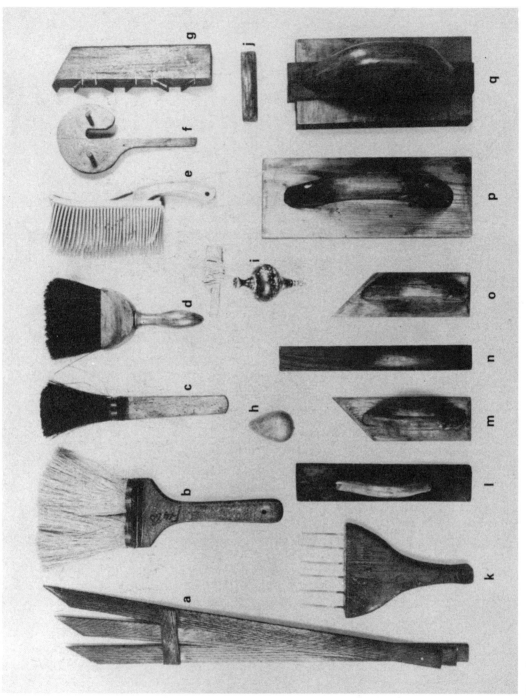

Plate 1

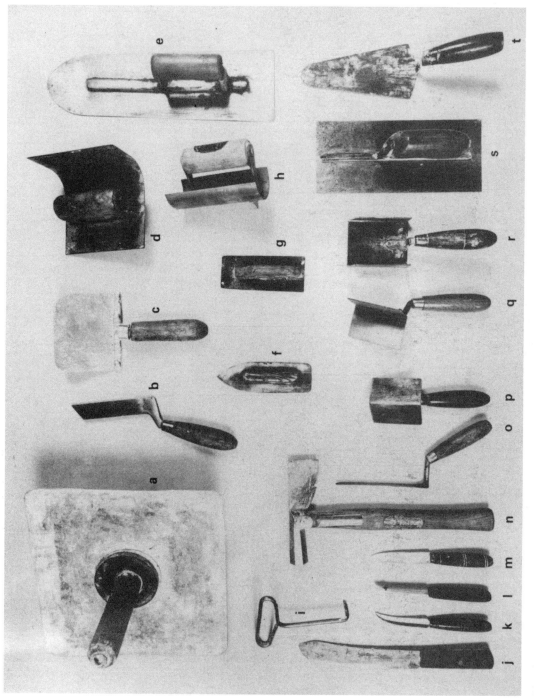

Plate 2

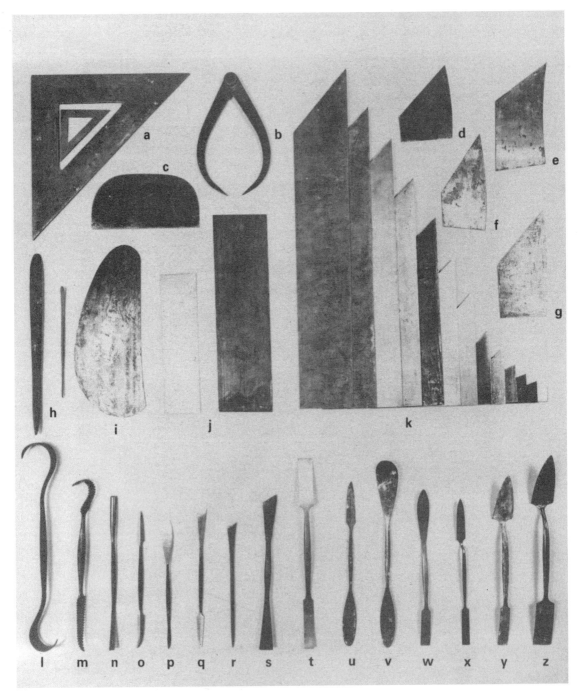

Plate 3

Plate 4

Hair hook – an obsolete appliance used to rake hair into a haired mix

Hammer – brad, club, lath (2n), tack

Hawk / handhawk (2a)

Hook – to remove casts from moulds (2i)

Joint rules – curved (3d, e, f, g), straight (3k)

Jointer (2b)

Jointing sponge (4d)

Knife – canvas (2j), jelly (2m), joggle (?l), stanley, trimming (2k)

Larry – a traditional tool used for knocking up large batches of coarse stuff and when making lime putty

Lines – chalk, plumb, string

Modelling tools – box wood (3h)

Pincers

Planes – french (1 g), jack, block, surform (French planes are not commercially available and must be home-made)

Press – for pressing composition in reverse mould

Saw – hand for plaster and wood, keyhole, hacksaw

Scissors – for canvas, glass fibre and Kevlar

Scraper

Scratcher, comb, (1k), lath (1a), wire

Screwdrivers – slot, posidrive

Scribe

Small tools – chisel (3t), corner chisels (3s), double leaf (3u), gouge (3n), griffins (3l, m), leaf and square (3w), picking (3o, p, q, r), rasp, spatula (3v), trowel and square (3x, y, z)

Snips / tinshears

Spirit level

Square – adjustable, radiating – normals to curves (1f), set (3a), shoulder

'Stickit' sander

Tape measure

Template, metal, adjustable (4a)

Topcutters

Trowel – angle, external (2q), internal (2p, r), coving (2d, h), dabbing, dashing (2c), floor-laying (2e), gauging (2t), laying-on (2s), margin (2g), margin, pointed (2f), internal angle (2p).

TOOLS, CARTRIDGE-FIXING

These are usually operated by hitting with a hammer or depressing or pulling a trigger. A nail, complete with metal or plastic washer, is placed in the firing position, and a cartridge loaded into the breach. Goggles must be worn when operating these tools, and every safety precaution listed in the manufacturer's instruction adhered to.

Many makers now issue a certificate of proficiency after a short course of instruction, and this is essential for safety.

Many items may be fired, from masonry pins to bolts and threaded studs. Cartridges are available in four grades: black–extra-high power, red–medium-high, yellow–medium-low, green–low. Can be fired into brick, concrete and steel members/beams. Fixing appliances can be fired into brick, concrete and steel members.

TOPPING

See BENCH; FLOOR LAYING AND ROOF SCREEDING.

TORUS

See MOULDING, SECTIONS.

TOW

See HAIR.

TRAMMEL

See COLUMN; ELLIPSE; ENTASIS; RUNNING CIRCULAR CURVES.

TRAVERSING

A floating rule between two SCREEDS – to rule off FLOATING.

TREAD

See STAIRCASE WORK IN GRANITE AND CEMENT.

TRIGLYPH

A block with three vertical channels (GLYPHS) cut in. It forms a distinguishing feature in the frieze of the Doric order of architecture. The channels are usually V-shaped in Roman and semi-elliptical in Grecian examples.

TRUSS

A rectangular block, enriched or plain moulded, acting as a support to a feature. Fixed with its depth exceeding its projection. For model see BRACKETS.

TURNING

The process in which the setting plaster is revolved on a spindle against a fixed metal PROFILE in a turning box. (See Fig. 124.) The centre of the spindle will be the centre line of the turned piece. The profile is therefore cut smaller to allow for the radius of the rod, and the centre of the rod is lined up with the profile. The rod may be held by wooden blocks or by passing it through a drilled metal bar. A metal stop prevents lateral movement if the rod is continuously pushed against it during turning. The rod may be keyed with a hacksaw or a file, or by passing nails through drilled holes.

If the turned piece is to be removed from the rod a tapered spindle is formed in plaster against a straight profile. A key is cut along this to prevent the turned piece slipping round it when it is shellacked and greased. MUFFLES are used to core out the piece if necessary. Compared with running a section up with a running mould, it is generally more difficult to obtain a finish. As with SPINNING, fresh plaster is applied and subsequently removed from the profile when, due to its stiffening, it ceases to roll.

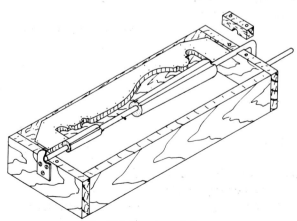

Fig. 124. Turning a baluster

Turning a spiral

The box is made up and works on the same principle as normal turning, with the addition of a drum on the spindle and a track to take a sliding stock. A rule overlapping the stock at the back of the track will hold the stock down to counteract the pressure of turning. The hinges will allow the stock to be lifted clear of the turned work when the stock is returned to the front for each successive run. The circumference of the drum must be the same as the pitch of the spiral. The wire (of steel brake or clutch cable to eliminate stretching) is fed through a pulley or eyelet from the drum to the stock and, as the spindle rotates, the profile is pulled along the section. The leading edge of the stock will need to be feathered away at a great angle as the angle of the spiral being turned will pass diagonally underneath it. The muffled run will show how much clearance is required (Fig. 125).

Turning multi-sided figures

Figure 126 shows how the sides, along with the mitres of a multi-sided figure may be run up with a running mould by turning the figure in the box after each successive run. See MITRE, RUNNING UP/ MULTIPLE RUNNING.

TWIN SLIPPER

A running mould with a pair of slippers, a suitable distance apart, on the same side of the stock. Their function is to support the stock where there is no surface on which the nib may bear. (See Fig. 62.) See also MOULD, RUNNING.

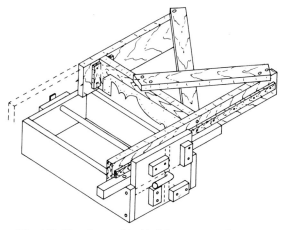

Fig. 126. Turning multi-sided figures: running up a four-sided baluster

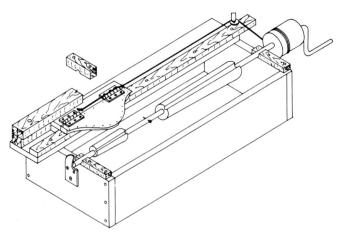

Fig. 125. Turning a special column

TYMPANUM

The triangular centre section of a PEDIMENT, enclosed by the horizontal and two raking mouldings.

TYROLEAN FINISH

See FINISH/RENDERING, EXTERNAL

UNBONDED FLOOR

See FLOOR LAYING and ROOF SCREEDING.

UNDERCOAT

Any plastering coat other than the finishing: pricking-up, render coat or scratch coat; floating coat.

UNDERCUT

The overhang in a model or mould which will prevent a cast from being removed from it. See MOULD, REVERSE.

UNSOUND

See FAULTS.

VAULTING

A ceiling constructed of a single, continuous arch or a series of intersecting arches.

Fibrous

Barrel vaulting

A curved ceiling, formed like a single continuous arch and commonly known as 'barrel ceiling'. A term usually applied to a barrel semi-circular in section but sometimes having a prefix to describe other types of curve, e.g. elliptical barrel, segmental barrel, etc.

Moulds for this type of work are DRUMS that will produce rectangular PLAINFACE of the correct curve to make up the ceiling. The drum is formed using RIBS cast to the required curve.

Casts are cast as plainface with variations to accommodate the type of curve and the method of fixing. It is feasible with small ceilings to fix to timber bearers running lengthways, in which case all laths running either way in the cast are laid flat as in a plainface to be fixed to timber, those round

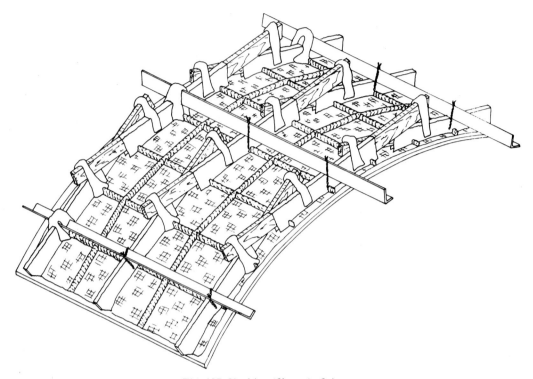

Fig. 127. Vaulting (fibrous): fixing

the curve being BRUISED if the curve is sufficiently quick. It is more usual, however, to fix such ceilings to metal bearers. This may be done by reinforcing with 6 mm lath on edge, those round the curve being broken with knuckle joints, the length of the laths depending on the quickness of the curve.

The casts may be fixed more quickly by on-edge integral 25 mm × 50/75 mm timber battens laid break joint round the curve and positioned to suit the metal fixings. In this type of fixing, less metal and fewer fixings are required. The battens make the cast semi-self-supporting so that only suitably suspended 75 mm angle irons are needed at 1.200 mm centres. When a curve is too quick for normal casting rules they may be built up by nailing 6 mm wood laths one on top of another. Provision is made for a suitable rebate to receive scrim in the

stopping, and a lapped and rebated joint will be found a great aid when fixing.

Fixing the casts is as for PLAINFACE. Where integral battens are used the fixing wires may be passed under them without piercing the face of the cast (Fig. 127).

Stopping is as for PLAINFACE, using a straight joint rule on the joints running over the curve and a curved joint rule or busk on those running along the barrel.

Vaults

A term usually applied to a ceiling formed by a series of cross-vaults intersecting at right angles with one or more main vaults, all of the same height although

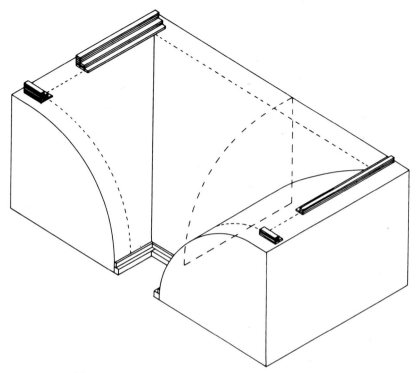

Fig. 128. Vaulting (fibrous): drum to cast cross-vaulting

the main vaults may be wider. (An elliptical section gives a greater span.) With multiple main vaults and the cross-vaults spaced a small distance apart, squares will be formed at the bottom of each inter-section where columns or piers are positioned as supports. These may be functional, as load-bearing concrete, brick piers or steel stanchions with metal-lathed cradling, or false, as metal lathed cradling or solid or fibrous COLUMNS.

Reverse moulds for this type of vaulting consist of a DRUM made up of two drums intersecting at right angles to form the groin line. Casting rules are used to form the joints between the casts across the barrels. Where the curve is too quick to bend a 25 × 50 mm rule, the rules are made up by fixing 6 mm wood laths, one on top of another to the correct height. Floating and setting rebates may be needed

where the bottom picks up with solid work. Lapped, rebated joints aid the fixing and pick-ups from one cast to another. Where the barrel is sufficiently small it may, of course, be cast in one, with the cast repro-ducing half of the other barrel, but usually the curves are sufficiently large for the cast to produce the groin line and half of both barrels, needing four casts to make up each intersection.

Very large sections require a drum to be made up as described but manageable casts are taken from it to make up the unit in sections. The drums are constructed of RIBS of the correct shape set up as shown in Fig. 128. An accurate groin line may more easily be formed over the finished surface of one drum by ruling it in as a screed from the ribs of the other drum. If necessary, this line may then be worked on to make it true. It is then shellacked and

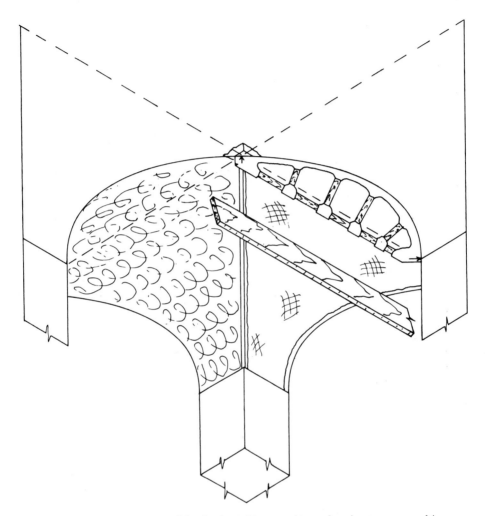

Fig. 129. Vaulting (solid): floating with pressed-screed groins to cross-vaulting

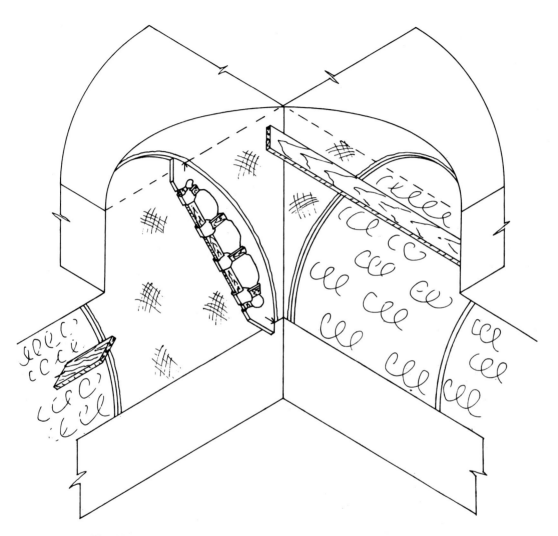

Fig. 130. Vaulting (solid): floating with pressed screeds to intersecting vaults

greased and used as an extra rib when ruling in the second drum.

Casts forming this sort of work are made to be self-supporting with integral 25 mm × 50/75 mm timber battens which are also used as their fixings. Such casts tend to be the largest produced in fibrous plasterwork as an attempt is made to produce the groin line in one piece. They are cast in the same way as curved PLAINFACE.

Fixing is effected at the top and bottom of each cast. Where the groin line is cast in one the bottom of the cast is fixed to the wall or pier. This is done by nailing or screwing to timber battens on solid backgrounds and by wire and wad round the angle iron that runs along all barrels. On large vaulting, where the groins need to be made up of more than one cast, extra fixings will have to be provided at the intersection of each cast.

Stopping is as for barrel vaulting. See also LUNETTE.

Fan vaulting

Where all RIBS have the same curve.

Ribbed vaulting

When mouldings cover the groin lines. Figure 132 shows a running mould in position to run a moulding over a groin, either as a model or run *in situ*. The mould is of the three-legged half-slipper type to run round the curving groin line over a curved vault. The two-piece stock is pivoted at the centre as shown, to allow the profile to open – from a right angle at the springing line to an obtuse angle as it runs towards the top, and close to a right angle again as it travels down the other side. The pressure applied when running is to open the mould, keeping it against both running rules. The rules are positioned to the running mould held on to the setting out as in PEG MOULD; RUNNING CHANGING CURVES.

Vaulting pendentives

Figures 133 and 134 show the DRUMS for casting PENDENTIVES.

Solid

Barrel vaulting

Figure 129 shows the intersection of two barrel ceilings, being ruled in from pressed screeds. The groin lines are formed as MITRES. For method see FLOATING TO CURVED BACKGROUNDS.

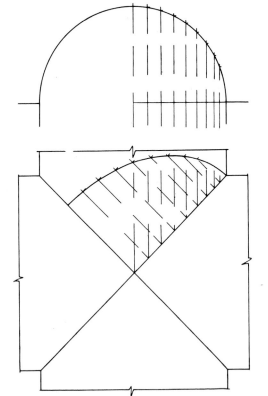

Fig. 131. Vaulting (geometry): to plot the section at the groin of intersecting vaults

Vaults

Figure 129 shows the curved surface formed by the intersecting vaults; this is being ruled in from the pressed screeds of the groin.

Points on the vault are dropped on to the plan of the groin line and projected off, perpendicular to it

(Fig. 131). The height of each point from the springing line in the elevation is marked on the corresponding ordinate off the plan of the groin line to plot the curve of the rib.

VERMICULATION

See FINISH/RENDERING, EXTERNAL.

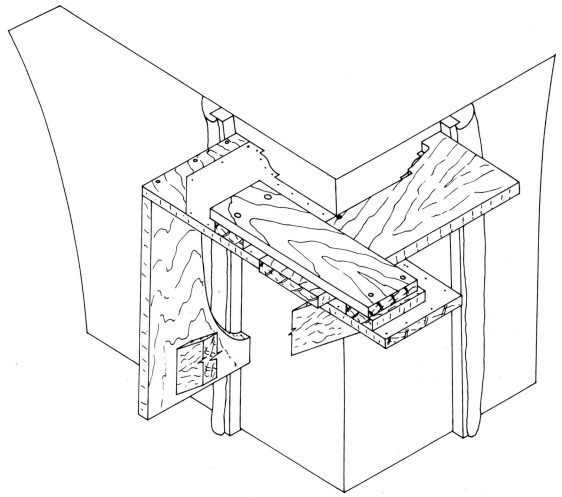

Fig. 132. Vaulting, ribbed: running a rib over a groin

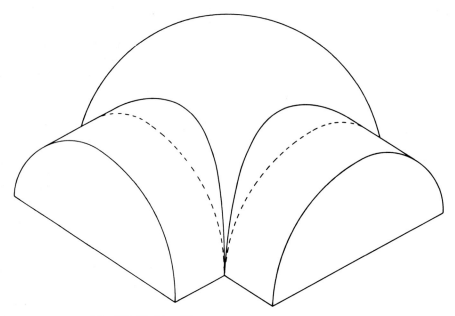

Fig. 133. Vaulting (fibrous): a drum to cast a pendentive

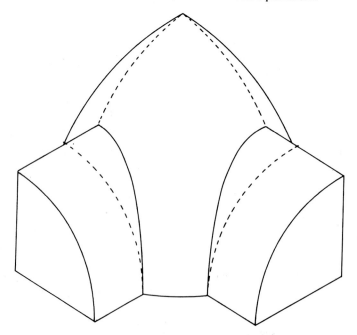

Fig. 134. Vaulting (fibrous): a drum to cast a pendentive

VERMICULITE

See AGGREGATE.

VIBRATOR

See MECHANICAL PLASTERING.

VOLUTE

The spiral feature found in Ionic, Corinthian and composite capitals.

WADS

Canvas dipped in plaster of Paris until it is completely impregnated. Used for fixing units together – in CASTING, attaching battens to backs of casts; in FIXING, joining one cast to another and casts to the fixings. See also LAPS.

WALLBOARD

See BACKGROUND.

WASTE MOULD

See under MOULD, REVERSE.

WATER LEVEL

This appliance consists of two capped glass tubes, known as sighting tubes, connected by about 20 m of rubber tubing.

It is filled with water until the water level shows half-full in both sighting tubes – the entire system is based on the principle that water will find its own level. The water level is used to transfer level points from one part of a building to another and is most convenient to use as it can travel around, under or over almost any obstacle.

To operate, one holds or hangs sighting tube 1 so that the starting point will come halfway up the tube. Sighting tube 2 is then taken to the place where the second point is required. Both caps are now opened slightly and either sighting tube is raised or lowered until the liquid in tube 1 shows the same level as the starting point. The surface of the liquid in tube 2 will then show exactly the same level as the starting point.

It is important to note that both ends must be opened slightly and that the tubing must be free from knots and kinks.

WATERPROOFING

Types of operation

1 To provide a dense impervious skin to the outside of a building.
2 To provide a water-resistant coating to the inside of a specific part of a building, e.g. a basement.

Normal cement–sand mixes, while offering a degree of resistance to water penetration are inadequate for full waterproofing. Therefore, for this type of work an ADDITIVE either in powder or liquid form is essential. One of these should be added to the mix and distributed throughout it thoroughly.

When in powder form the compound acts as a lubricant to the cement and sand. It fills in all the pores and voids that normally exist between the grains. In this way it builds up an extremely dense mass. One disadvantage is that the powder is difficult to mix with the other materials. For the best results, it should be mixed dry with the Portland cement, passed through a fine sieve and then mixed with the sand. One way of eliminating this problem is to use ready-mixed water-resistant Portland cement.

The other compounds are either in a liquid form or as a paste. In both cases they are mixed in with clean gauging water. These will combine with the cement and cause a more complete crystallisation.

No lime whatsoever should be added to mixes that are to be used for waterproofing. The sand must be clean and well graded and no knocking back should occur. In all cases, when using a waterproofing compound, the manufacturer's instructions must be adhered to.

Another method used to protect existing renderings from water penetration is to apply a coat of colourless liquid waterproofer by either brush or spray. These do not affect the final appearance of the finish. Several are obtainable and are basically compounds of selected silicon resins in hydrocarbon solvent.

External

The background for this must be clean and all joints well raked. At the time of the initial application it should be kept damp so as to prevent DRY-OUT. Where necessary, because of the lack of adequate key, a SPATTERDASH coat of 1:2 cement:sand, plus either the waterproofing compound or a purpose-made waterproofed cement bonder should be applied to the entire area.

All work must be carried out in a minimum of two coats and contain no joints. The first coat of 1:3 cement:sand plus waterproofing compound is applied to an average thickness of 9 to 10 mm. When this is complete it must be well keyed to receive the next coat. The keying is carried out by a wire or comb scratcher, but must not penetrate through to the background. This coat is then left for two to three days, during which time it must be kept damp, either by spraying, covering with wet sacks or by sheeting in heavy PVC.

For a plain finish in a position of extreme exposure the same mix of cement, sand and waterproofing compound may be used. For the final coat when a textured finish is required (see FINISH/RENDERING, EXTERNAL), the mix will depend upon the type of finish required, and the degree of exposure. Where any doubt exists concerning this, a second waterproofed coat should be applied.

Internal

Very often, in the case of basements this will mean carrying out this type of work to all walls, or at least external walls and also the floor surface. Possibly the most important factor here will be the angles between walls and between walls and floor. As the work should be in two coats the best approach is to stagger all internal angle work and dub out to form small coves.

For the walls, a waterproofed spatterdash coat is generally essential, then two coats of cement:sand, 1:2 or 3, plus waterproofer – one as a render coat, the other as a floating coat. Under certain conditions, this floating coat may also provide the finish. If this is to be, a dry mix will be required as well as a wet one. As there is no suction from a waterproofed backing the driers will be applied to the second coat after it has been ruled off straight so as to allow it to be rubbed up with a wooden float.

The floor may be either a 1:2 cement:sharp sand, plus waterproofer floor SCREED, or 1:2 cement, granite and waterproofer floor TOPPING. In both cases the floors should continue for a distance of 25 to 50 mm up on to the wall. Once it is firm, a rebate should be cut in this, back down to the floor so that the second application's grout joint of wall plastering will first overlap the related section and then eventually finish flush with the floor finish.

No joints should occur anywhere but if they are unavoidable they must be staggered, splayed and grouted. Dubbing out to all internal angles should follow the spatterdash coat. Where a gypsum plaster finish is specified it will be either Class B, C or D plaster, neat or with up to 20% lime putty. Condensation may occur but this can be overcome by good ventilation.

There is no definite rule for the order of working and each job should be decided on its merits. Points to note are:
1　Where the water penetration is at its worst.
2　The length of time the job will take.
3　The floor must be left for several days for hardening and curing.

All electrical work, plumbing and similar operations should be completed before waterproofing begins.

WATTLE AND DAUB

Wattle

Wooden rods, usually hazel, woven in and out of upright slats of oak.

Daub

The working-in by hand of mud or clay, toughened with straw, so that all the wattles are covered both inside and out. This method of construction dates back many hundreds of years.

WAX, MOULDING

Moulding wax is made up by melting together equal quantities of beeswax and rosin (a natural resin in block or powder form) over a low flame. Overheating will cause the wax to become brittle. Extra quantities of either constituent may be added if the mixture proves too brittle or too soft. Wax is particularly good for reproducing small casts that would prove extremely difficult to release from plaster moulds without splaying all vertical surfaces and attaching elaborate lifting-out sticks. Although classed as a rigid moulding material, it may be softened slightly with warm water so that the mould can be bent away from the cast, allowing vertical and even slightly undercut members to be present on enrichment. As the material is basically brittle and cannot be reinforced, use is limited to small flood moulds.

Seasoning the model

Damp clay models need no preparation. Plaster models are saturated by submersion in water and their surfaces made glistening moist by stippling off or brushing on water immediately before pouring.

Clay is the most convenient material for retaining walls. The wax is removed from the heat and, provided no signs of overheating are present, tested by thoroughly wetting a finger in clean water and dipping it into the wax and immediately back into the water. The experienced plasterer can thus judge the temperature both by 'feel' and by the test mould formed on his finger – too hot and it will stick, too cool and it will wrinkle. The molten wax should be poured so that it flows across the face of the mould steadily and continuously. Should the flow cease, a seam will result in the finished mould, causing the mould to break as glass does when it is scored.

Seasoning the mould

When cold, the mould is removed from the model and washed out with soft soap or a clay slurry with a little tallow. It is ready for use when dried and is lightly oiled before each casting operation.

273

To remove a thin flush cast from a wax mould the whole is held in the hand with the back of the cast uppermost and the mould strike-offs bent back slightly from the cast by pressure from the thumbs. If this does not release the cast, the whole may be immersed in cold or warm water. Capillarity will draw the water between mould and cast; if the water is warm it will make the mould more pliable.

WEATHERING

See MOULDING, SECTIONS.

WHITING

White chalk ground to a fine powder, used in the stiffening of CARTON PIERRE and COMPOSITION.

WOOD LATH

See BACKGROUND.

WOOD-WOOL SLAB

See BACKGROUND.

X-RAY PLASTER

This is an undercoat plaster for use on suitable backgrounds where protection from X-rays is needed. It is a retarded hemihydrate, premixed gypsum plaster which has barytes aggregate incorporated as an X-ray resistant substance.

Suitable backgrounds are walls of open-textured dense concrete blocks or keyed surface bricks, block or brick walls reinforced with rib-lath, anid concrete lintels reinforced with EML. For these the recommended maximum undercoat thickness is 25 mm. For solid soffits and timber joist floors, both reinforced with rib-lath, the recommended thickness is 15 mm. The setting time is 1.5 to 2 hours. When plastering over rib-lath the thickness of plaster should be measured from the face of the lath.

The finishing coat can be Thistle Multi-Finish or Thistle Board Finish.

ZINC PROFILE

See PROFILE, METAL.

APPENDICES

APPENDIX 1

SAFETY AT WORK

The building industry has no need to be as dangerous as it is. Since the Health and Safety at Work Act in 1974, regulations have been introduced to cover all the common hazards on site and in the workshop. These regulations are listed at the end of this Appendix. In something as diverse as the construction industry it cannot be easy to be sure that all the regulations are followed meticulously or even known about. The self-employed plasterer, the two-person firm, even larger enterprises, work on from day to day, perhaps unaware that the substances they are using or the scaffolding they are climbing on can be a threat to their lives. And does anyone really want to develop dermatitis or industrial asthma? Most hazards, right down to those causing minor skin irritation, can be avoided with the right precautions. It is up to everyone to make sure that the right precautions are taken *all the time*.

CONTROL OF SUBSTANCES HAZARDOUS TO HEALTH (COSHH) REGULATIONS

COSHH regulations follow on from the Health and Safety at Work Act 1974 which requires manufacturers and suppliers of substances for use at work to make sure as far as they possibly can that those substances will not damage the workers who use them.

Damage may be:

to skin – from mild irritation, through defatting, to dermatitis

to eyes – from mild irritation to permanent damage
by inhalation – drowsiness, headache, sickness
by swallowing – sickness, pain.

Some precautions are a matter of common sense. Avoid substance contact with skin and eyes. Wear appropriate protective clothing. Use barrier cream. Don't eat, drink or smoke while working and always wash your hands before eating, drinking or smoking.

Always read the labels on containers.

As with all safety measures, your employer must provide proper protection but the worker at the bench or on site has a responsibility to make sure that he/she has taken every measure for self-protection. Manufacturers issue COSHH sheets for each substance you are likely to use and you should read (and absorb) the COSHH sheet for every new material before starting to work with it. If you have been working with a material but have never read its COSHH sheet, make sure you get hold of a copy before you continue. All employers must make themselves familiar with the detailed COSHH requirements.

If you read through the list of COSHH sheet contents that follows under 'The Employer's Duties', you will see how extensive they are *and* when you realise that each heading may have many subheads you will comprehend that it is impossible to cover them fully here. We can only offer an indication of things you should be especially aware of.

Fire hazards

As well as hazards directly to one's person, many substances also present a fire hazard and those with a flashpoint of less than 32°C must be stored and handled carefully. No more than 50 litres may be stored in a workshop and these 50 litres must be kept in closed vessels within a fire-resisting cupboard or bin. Don't keep flammable materials in a plastic bucket or any other container that is not fireproof. Full requirements are given in the Highly Flammable Liquids and Liquefied Petroleum Gases Regulations 1972.

Tidiness and a clean working environment contribute enormously to fire safety:

- No naked flames
- Empty bins with hazardous waste daily and keep them damped with water on hot days.
- Don't leave rags soaked with solvent lying around.
- Cans can explode when empty. Dispose of them immediately.
- Always reseal part-used cans while in use.
- Return cans to storage when not in use.
- Peroxide containers must be disposed of separately.
- A fire extinguisher should be by each fire exit. Use one if there is a small fire – not a bucket of water.
- Escape routes and exits must be kept clear. Exits must never be locked when people are at work.

There should be a plan of action for the outbreak of fire and everyone should know what to do. If a fire does break out raise the alarm immediately, by shouting 'Fire!' if no other alarm system is immediately to hand. Anyone near the alarm can then set it off. Make sure everyone hears and then you must all leave the area. You may have to use a fire extinguisher to allow you to escape but serious fires are the job of the fire brigade. Call them.

Specific hazards

There is no substitute for being fully aware of the contents of COSHH sheets but we would draw special attention to these particular hazards and precautions they describe.

Accelerators (cobalt solution) *for polyester resins*: have a flashpoint of 31°C and storage regulations must be

adhered to. Never let a neat accelerator come into contact with a neat catalyst; at best there will be a fire and at worst an explosion.

Acetone: Plasterers usually use acetone only for washing brushes and laminating tools but this must be done while wearing goggles and butyl rubber gloves. *Never use acetone or any other solvent for cleaning your hands.* It irritates eyes, skin and the respiratory system. Prolonged exposure can cause drowsiness, nausea, dizziness, headache and uncoordination. If it is inhaled or ingested unconsciousness can result. If this happens, keep the affected person away from heat, sparks and flames. Move him/her to the fresh air immediately and get prompt medical attention. Acetone is highly flammable with a flashpoint of –18°C and must be stored with great care in the recommended way. A metal flip-lid, treadle-operated bin is the best container for acetone when it is to be used for washing out, etc.

Catalysts (organic peroxide) *for polyester resins*: These are very poisonous if swallowed and are irritating to skin and eyes. Use gloves, goggles, protective overalls and a fume respirator. Always ensure entire contents of a container have been used and reseal empty containers before disposing of them. Do not let peroxide catalysts come into contact with other substances, especially accelerators. Store them separately below 30°C. Never pour unused material back into its original container. Only the minimum amount for current use should be kept in the workshop. Keep catalysts from contaminating rags and clothing because this provides a danger of spontaneous combustion.

Catalysts (organo tin) *for silicon rubber*: These are harmful if inhaled, absorbed through skin or swallowed. They cause mild skin irritation but severe eye irritation. They should only be used in a well ventilated area. Wear disposable gloves, barrier cream and safety glasses. Do not let organo tin catalysts come into contact with water, a heat source or flame, acids and oxidising agents.

Cement: When cement is mixed with water or becomes damp a strong alkaline solution is produced. This can cause serious burns or ulceration to the eyes or skin. Before they set, cement mortar and concrete mixes can cause both irritant and allergic contact dermatitis. Use dust masks, gloves and goggles. For complete protection when laying concrete, use waterproof gloves, trousers and wellington boots.

Glass fibre reinforcement: Particles in the eyes will cause irritation and some workers will find their skin is irritated. It is claimed that there are no long-term effects from inhalation but breathing in high concentrations of airborne dust may cause irritation of the upper respiratory tract. Use barrier cream, long sleeves, dust masks and goggles.

Special note: Abrading of the set resin laminate by sawing, sanding or drilling may produce particles that are harmful but not covered by the COSHH sheets on either Glass Fibre or Polyester Resins. Hooded overalls must be worn, together with goggles and a dust mask to cover the whole of the face.

Lime, hydrated: The main constituent of hydrated lime is calcium hydroxide which irritates skin and eyes and can cause burns in the presence of moisture. Prolonged contact with lime can cause dermatitis. Use gloves, overalls and goggles. If some gets in your eyes you must wash them out for at least 20 minutes and seek medical attention immediately.

Polyester resins: Flashpoint 23–34°C. The significant constituent of polyester resins is styrene. Long-term carcinogenic effects have been suspected but not proved. If you already have chronic respiratory or skin disease or some disorder of the central nervous system, styrene may aggravate it. It is classed as moderately irritating to skin and eyes, slightly toxic if inhaled and moderately toxic if ingested. It should only be used in a well-ventilated area. Wear safety glasses or chemical goggles, impervious gloves and overalls. *Do not use contact lenses when working with polyester resins.* It is standard practice in the USA to earth electrically all containers and pipes and this is recommended by one polyester resin manufacturer in the UK under 'General Precautions' on their COSHH sheet.

Machining castings will cause styrene fumes and resin dust, which should be kept as low as possible. Use all the precautions as for *Glass fibre reinforcement* above.

Polyurethane foams: These come as a two-pack item. One of the packs contains isocyanates which are particularly dangerous as they can cause industrial asthma or even death if inhaled extensively. A combined carbon filter and particulate filter should give medium-term protection but, for real safety, air-fed masks must be used, together with gloves, barrier cream on exposed areas of skin and a face visor.

Polyvinyl alcohol release agent: Over-exposure irritates the eyes and repiratory system. Excessive concentrations may cause drowsiness or even loss of consciousness. If either of these happens, remove the patient to fresh air immediately.

Dust: Dust inhalation can damage the repiratory system. It can cause:

running nose and eyes
sore throat
headache and dizziness
inflammation of the lungs, leading to bronchitis.

Eyes can be damaged, too. When workers are in a dust-laden atmosphere respirators must be provided, together with suitable protection for the eyes. If possible, a wet-abrading process should be used as this does not release dust. Always damp down dust-covered areas with a fine spray of water before sweeping, unless the dust consists largely of hydrated lime which is harmful when wet. In manufacturing areas an industrial vacuum cleaner should be used to remove dust that has not been removed by the extractor.

Gypsum plasters: Plasters are not classified as hazardous but may irritate eyes or sensitive skin. Dust can irritate your repiratory system. Work areas should be ventilated and dust controlled. If these precautions are not possible, wear a half-face dust mask. Plasters used in solid plaster work are alkaline and have a more detrimental effect on personal contact than most of those that are used in fibrous plasterwork. Plasterers' grease is usually sufficient protection against standard casting plasters that are acidic, but gloves and barrier creams should be used during long casting operations where the grease tends to wear off, or with those that are alkaline, and by those workers who are sensitive to all forms of plaster.

First aid

It would be dangerous to try to offer first-aid advice in a book such as this. The Health and Safety (First Aid) Regulations suggest (but do not insist) that there should be a first-aider for every 50 people employed. If there are particular risks associated with a material or working practice the employer must ensure that first-aiders receive training to deal with anything resulting from those risks. Details of the first aid specific to a material are always given on its container, even if the COSHH sheet is not immediately to hand, and should be followed precisely in cases of emergency. In general:

- Make sure the first aid cabinet is handy and well-stocked.
- Make sure eyewash bottles are easily accessible. Washed affected eyes for at least 15 minutes.
- Make sure there is a good supply of goggles, masks and disposable gloves.

THE EMPLOYER'S DUTIES

A safety policy

Under Section 2(3) of the Health and Safety at Work Act 1974, everyone who employs five or more people has a legal obligation to make a written statement of their health and safety policy. The statement must be their own, specific to their enterprise, setting out the general policy for protecting the health and safety of employees at work. It must show how the policy is put into practice. It must be brought to the attention of all employees who should be able to read, understand and follow it. Though the main responsibility for health and safety lies with the employer, employees at every level must carry out the provisions of the policy and should be able to see how they fit into the system. They must know where to go for advice, to report an accident or to get help.

Before a policy can be compiled risks must be assessed:

- identify the hazards
- decide who might be harmed and how
- evaluate the risks arising from the hazards and assess existing precautions. Are they sufficient or should more be taken?
- record the findings.

Employers cannot compile a safety policy and sit back. Safety is an on-going thing. They must ensure, by delegation if necessary, that all equipment is maintained in a safe condition and that personal protective equipment such as helmets, boots, goggles and respirators are available and used when needed. Hazardous materials must be stored and handled correctly. If there is too much noise in the workplace ear protectors must be provided. Building workers coming new to a site may encounter unexpected hazards unless care has been taken to avoid these by consultation between the employer and the site manager.

All fire exits must be marked, unlocked and free from obstruction. Fire extinguishers must be provided and fire drill practised. Someone must be responsible for first aid and everyone should know where the (well stocked) first-aid box is.

The employer must give the workers all necessary information about the materials and machinery they are using. Employees, supervisors and managers must be trained so that they can carry out their work safely. They must be well supervised so that anyone new to the work does not run into avoidable danger.

The safety policy must be regularly monitored, assessed and revised in the light of experience.

COSHH

Under the Health and Safety at Work Act, manufacturers and suppliers of substances for use at work have a responsibility to ensure, as far as is reasonably practical, that they will have no risk to health when properly used. Suppliers are also required to provide users with information on the nature of any hazards of the material and recommend precautions for their safe use. This information is provided in the form of COSHH (Control of Substances Hazardous to Health) sheets which detail:

(1) Identification of the substance/preparation and of the company;
(2) Composition/information on ingredients;
(3) Hazards identification;
(4) First aid measures;
(5) Fire-fighting measures;
(6) Accidental release measures;
(7) Handling and storage;
(8) Exposure controls/personal protection;
(9) Physical and chemical properties;
(10) Stability and reactivity;
(11) Toxicological information;
(12) Ecological information;
(13) Disposal considerations;
(14) Transport information.

The employer has a responsibility for establishing safe working practices with the materials. This responsibility includes *asking the supplier* for specific information on substances the workers are being required to use and passing that information on to the workers. The workers, in turn, have a duty to use the safe working practices recommended. Where there are particular risks associated with materials or modes of work the employer must ensure that first-aiders receive training to deal with these risks.

Wherever it is needed, employers must supply their workers with suitable personal protective equipment and make sure that it is properly used. For more details see the section on personal protective equipment on p. 283.

Construction (Design and Management) Regulations 1994

The CDM Regulations are intended to improve the overall management and coordination of health, safety and welfare throughout all stages of a construction project and place a duty on everyone who can contribute to this. The client must appoint a principal contractor to coordinate and manage health and safety during construction and a planning supervisor to perform the same function during the design and early stages of preparation. Except in very small projects, the planning supervisor may not be a single individual but an organisation, such as an architectural practice or a company of consulting engineers, with experience of design, construction and health and safety. The planning supervisor's duties are to:

- Notify the Health and Safety Executive of the project;
- Establish cooperation between the designers and ensure that they comply with their duties;

- Make sure that a pre-tender health and safety plan is prepared by the principal contractor;
- When asked, give advice to the client on such things as the competency of designers and contractors and the suitability of the health and safety plan;
- Make sure that a health and safety file is prepared.

Projects that are exempt from the CDM Regulations are: construction work that does not last longer than 30 days and does not involve more than four people; construction work for a domestic client; construction work in such places as offices and shops where ordinary work goes on as usual.

Welfare facilities

Away from permanent premises, eg on construction sites, the welfare facilities that should be provided *or arranged for* include toilets, washing facilities, changing and rest areas, drinking water and eating facilities. At the very minimum, clean lavatories and hand-washing facilities with warm water must be available. If the work is going to last no more than a week, arrangements can be made to use private facilities, eg in a cafe. The owner should give his permission and workers must be informed of the arrangement.

The **Health and Safety Executive** is the fount of all wisdom on these matters. It provides free leaflets on every aspect of health and safety and also publishes an extensive range of books. HSE's InfoLine 0541 545500.

ELECTRICAL SAFETY

Electricity must be treated with respect. In a work situation familiarity can lead to carelessness and even downright foolhardiness, such as thinking that a plug is not really necessary or that rubber-soled footwear can protect you in all circumstances.

There are three ways in which electricity is a hazard. The obvious one is contact with live parts. This will cause shock, burns and even death. The second way is a fault which may cause a fire. The third is as a source of ignition in a potentially flammable or explosive atmosphere; a spark in dust from combustible materials can cause explosions. You should always be very aware of the greater risks in wet surroundings, out of doors (where equipment may become wet), and in cramped spaces with earthed metalwork where any fault could cause a shock.

Extension leads are apt to become damaged, as are their plugs and sockets. Equipment which is moved around a lot may well develop unsafe electrical characteristics. Mobile equipment, such as cherry-pickers and forklift trucks, can run over sockets and plugs inadvertently and cause damage. Sockets and plugs are safer if kept off the floor, not just to save them from such damage, but also to keep them out of any puddles or water spillage. Trailing leads can easily be tripped over and this not only damages connections but may cause the electrical appliance being used to jerk dangerously in the user's hands. Check plugs for damage and never use damaged connections.

Employers should make sure that new electrical systems are installed to BS 7671 and maintained meticulously. Existing installations should also be properly maintained. There must be enough sockets. Overloading by using too many adaptors can cause fires.

- When using portable equipment, plug it into the nearest socket so that it can be easily disconnected in an emergency and people do not trip over it.
- Overhead supplies to benches are best.
- Make sure that the ends of flexible cables have the outer sheath firmly clamped to stop the wires pulling out of the terminals.
- Replace damaged sections of cables *completely*.

- Use proper connectors or couplers to join lengths of cable. Do not use connector blocks covered in insulating tape.
- Some types of equipment are double-insulated and marked with a 'double square' symbol. The supply leads have only two wires – live (brown) and neutral (blue). Make sure they are properly connected if the plug is not moulded on.
- Protect light bulbs.
- Only purpose-designed electrical equipment should be used in flammable or explosive atmospheres.
- Where possible, switch everything off before plugging in or unplugging.
- Never clean or adjust equipment if it is not switched off and unplugged.

Extra safety measures

Temporary lighting can be run at lower voltages than the norm and battery operated tools are electrically safe. Portable tools are available that are designed to run from a 110 volts centre-tapped-to-earth supply. A residual current device (RCD) can provide additional safety in equipment operating at 230 volts or higher. This is a device which detects some, but not all, faults in the electrical system and switches off the supply.

- Always use the RCD if one is provided.
- If it trips, look for the fault and ensure that it is rectified by a competent person before starting work again.
- If it keeps tripping and no fault is found, consult the RCD's manufacturer.
- Check regularly that the RCD mechanism is free and working by using its test button. Take faulty equipment out of use and label it 'Do Not Use'. Get it mended by a professional.

Water

Water conducts electricity freely and, being fluid, makes a perfect contact. Plasterers need to take special care.

- If your hands are wet don't switch on or off, plug or unplug, or hold electrical appliances.
- Don't stand in water.
- Don't allow water to come into contact with anything electrical.

The Institution of Electrical Engineers publishes a *Code of Practice for In-service Inspection and Testing of Electrical Equipment*, ISBN 0 85296 844 2.

PERSONAL PROTECTIVE EQUIPMENT (PPE)

Employers have a duty to supply their workers with PPE whenever it is needed. Self-employed plasterers, or those working alone, should check with the relevant COSHH sheet and make sure that they are properly protected. PPE includes hooded overalls, boots, gloves, masks and eye protection.

Construction boots

These may have slip-resistant soles and a steel mid-sole to guard against penetration by nails and a steel toe-cap to resist heavy objects falling from a height.

Gloves

These vary in weight and protectiveness. For plasterers' work the most commonly used are disposable PVC or rubber gloves that protect against irritation (and possible dermatitis) from solvents, catalysts, cement, glass fibre and even gypsum plasters themselves. Heavier gloves of cotton twill or leather protect against cuts and abrasions when manhandling materials in bulk and from burns during such operations as vinyl moulding.

Masks

A disposable paper mask or disposable cotton filter pad is usually considered sufficient to guard against large particles and non-toxic dust but there is a range available of moulded disposable dust masks (with and without valves) that appears to offer more safety. Some of these are proof against toxic dust or very fine dust, such as that from abraded glass fibre; a dust repirator with a replacement felt or impregnated wool filter can also be used. Non-toxic fumes are combated with a light fume respirator, consisting of a facepiece with a replaceable activated carbon filter. Breathing apparatus, which is a full-face mask connected to a filtered air supply, should be used when working with two-pack polyurethane foam. In all cases, check with the manufacturers' recommendations.

Safety helmets

Under the Construction (Head Protection) Regulations 1989, safety helmets must be provided and worn if there is risk of injury. They must fit, be comfortable and you must wear them properly, under instruction from whoever is in control of the site. The rules cover everyone on site, including all visitors. The only exceptions to this are Sikhs who wear turbans. There is a variety of designs and the most suitable should be selected, eg someone who needs unrestricted upward vision should pick a helmet without a peak.

Eye protection

The safest guards for eyes are all-round face masks that fit closely to the face and so protect your eyes from contaminants – splashes, fumes, dust particles – coming from any direction.

Protection for your hearing

Your sense of hearing can be damaged over a long period without your noticing until it is too late. Use ear muffs for harmful noises at a very high level and ear plugs or defenders for noise that is less but still constant.

GOOD HOUSEKEEPING SAFETY

Handling of materials in the plastering shop

For safety's sake, materials should be stored as in the next section on STORAGE.

Thoughtful use of a forklift truck can avoid much manhandling of bags of plaster. The truck can elevate the pallet, bringing each successive layer of bags to the top of the plaster bin where the bags can be rolled to the edge of the bin, slit, and emptied into the bin. About half of the dust in the atmosphere arises from this process; the other half comes from gauging up – the sprinkling of the plaster into the water causes particles to become airborne. A draught across the plaster bin, contrived or otherwise, can allow the operative to stand 'up-wind' and so out of the concentrated cloud of dust particles. Nevertheless, the dust is thus distributed to all parts of the plaster shop and it is far better to have the draught caused by an extractor which removes the dust from the atmosphere.

A forklift truck can also be used for demoulding and lifting heavy casts or moulds. If this is the intention it must be planned for from the start – the gangway must be wide enough (a forklift can move the benches) to facilitate manoeuvring and lifting-out sticks must be suitably placed for the use of slings.

Trolleys: A trolley is essentially a sturdy platform on wheels that is large enough to take a bucket and a bath. The trolley is the right height if plaster can be scooped without bending or overreaching from the bath or bucket on to the bench. A convenient design is an upside-down box with a castor-type wheel in each corner.

Skates: Another home-made device is a skate that is used for wheeling plainface-type casts on edge; it consists of a small wooden platform with a raised batten down each side to prevent the casts from sliding off sideways. One, in the middle of the cast, is sufficient for a cast of any size. Although several men may still be needed to keep the cast upright and under control, it is being wheeled and not carried.

Lifting

Always use lifting gear if you can. When you cannot, be careful to lift heavy things – sacks of plaster or large casts – in the correct way:

- Keep your back straight.
- Bend your knees.
- Tuck your chin in.
- Grip the object firmly with the whole of your hand.
- Hold your arms close to your body so that your body takes the weight.
- Stand up with a smooth movement.
- Don't be macho. If the object is really too heavy for you get someone else to help.

Gangways

Both working and access gangways should be kept clear of *rubbish*. Large items should be hauled immediately to outdoor bins or skips. Small amounts of rubbish should be put into locally placed bins which, in turn, must be checked frequently to see if they need emptying. *Spillages* in gangways – of water, liquid plaster, grease, oil, etc. – should be cleaned up immediately. Hazardous materials should be dealt with speedily, taking the precautions described in the relevant COSHH sheet.

Segregation

If at all possible, plasterers using different materials should be well segregated from each other. In this way the styrene vapours given off during fibreglass work, for instance, will be kept from workers not suitably masked.

Good practice

- Do not use a chisel instead of a screwdriver. A slipping chisel can result in damage to itself, the surface you are working on – or you.
- Do not leave nails only partially driven home in the expectation of easy removal later – a hand knocking against one can be cut to the bone.
- To prevent similar injury, a bench that is to have nails knocked halfway in to form a key should be greased *before* the nails are put in.
- When greasing wood *never* use a grease rag. Keep your fingers away from the timber by using a grease brush. A splinter picked up on a vigorous rub can be driven straight through the finger.
- Don't indulge in the common practice of burning off shellac surfaces – for all the very obvious reasons. And if your shellac brush should somehow catch fire, don't try to put it out in the shellac pot!

Slosh

Slosh is a slurry of dead plaster and water, resulting from washing out casting equipment – brushes, bowls, buckets, baths, hands – in a tank. The plaster is allowed to settle, water is removed from the top, and the concentrate saved to be taken away by commercial 'slosh-gobbling' tankers.

The amount of slosh produced can be reduced if surplus plaster is left in suitably greased containers and 'knocked out' when set. It can then be disposed of with the other solid rubbish.

The slosh tank system can vary from a basic, mobile, single, plastic container similar to a domestic water tank, to an elaborate construction of interconnecting pipes and series of elevated static tanks, designed to give maximum settlement.

Never eat or drink while at the bench in the plastering shop. Hygiene apart, there is always the possibility of absent-mindedly picking up a beaker of some noxious fluid rather than your cup of tea. Foodstuffs can easily become contaminated by toxic substances.

STORAGE OF MATERIALS

In the plastering shop

Gypsum plasters: 25 kg bags are supplied on pallets. Store them on the pallets, never on a concrete floor; keep them dry; stack them in such a way that forklift trucks have easy access. To avoid puncturing the bags, take all necessary precautions to protect from impact damage. Plaster will set if it comes into contact with air. Use bags in rotation and close opened bags between use.

Cement and lime: Treat these in exactly the same way as gypsum plasters. Like plaster, cement will set if exposed to air and, unlike plaster, cement and lime can both cause burns or irritation if damp.

Lime putty: Should be kept in a lidded bin to protect it from contamination and drying out. This is a dangerous material and a lockable container is desirable.

Sand: Should be kept in a clean condition, protected from all forms of contamination – salts, cats, etc.

Timber sections: Store in racks for stability and control.

Canvas: Is composed of fine, combustible fibres and easily catches alight. Store so that it cannot come into contact with any form of ignition.

Flammable materials: For safety precautions for highly flammable materials see the section on fire hazards on p. 277.

Shellac, waxes, etc.: These should be stored on shelves and clearly marked. Lids must be kept tightly closed to avoid evaporation and skinning.

Casts: Must be protected from damage and held to shape. *Cornice type casts* should be tied in pairs, face-to-face, as near upright as possible so that they are kept straight and safe and their faces are protected. Stand them against scaffold boards leaning against a wall or frame, or lay them horizontally on edge in a rack. *Flat casts*, such as plainface, are cleated face-to-face in pairs and cleated securely into a frame. *Contoured casts* should stand stably on their lifting-out sticks and braces so that these act as legs.

All other materials should be stored safely, off the ground, in a protected store.

On site

Plasterboards: Must be kept dry and in their wrappings until they are needed. They are taped together in pairs and should only be separated when they are about to be used. They should be stored on flat pallets that are large enough for the plasterboards not to overhang. If not close-boarded, the pallets should have slats not more than 450 mm apart. Plasterboards should be put in stacks of not more than three packs high. The platforms should be constructed to facilitate handling by forklift trucks.

When handling, do not drag one board over the other. Do not allow excessive bending of the boards: always carry boards on edge, using slings if necessary with spreaders to protect the board edges if they are being damaged by the slings.

Metal lath, beads, etc.: Avoid distortion. Do not damage the coating on metals. Keep them dry to prevent rusting.

Plasters, limes and cements: Treat as in the plastering shop, described above.

All other materials: To keep dry, store under cover and off the ground. Protect from damage. Valuable materials, along with site tools, are usually kept in a lockable metal site box.

LEGISLATION AND REGULATIONS

Health and Safety at Work Act 1974
Construction (Design and Management) Regulations 1994
Construction Regulations 1961–1966
Construction (Head Protection) Regulations 1989
Control of Substances Hazardous to Health Regulations 1994
Electrical Equipment (Safety) Regulations 1994
Health and Safety (Information for Employees) Regulations 1989
Highly Flammable Liquids and Liquefied Petroleum Gases Regulations 1972
Management of Health and Safety at Work Regulations 1992
Manual Handling Operations Regulations 1992
Noise at Work Regulations 1989
Personal Protective Equipment at Work Regulations 1992
Provision and Use of Work Equipment Regulations 1992
Reporting of Injuries, Diseases and Dangerous Occurrences Regulations 1995
Safety Representatives and Safety Committees Regulations 1977
Workplace (Health, Safety and Welfare) Regulations 1992

APPENDIX 2

USEFUL ADDRESSES

Akzo Nobel Chemicals Ltd
PO Box 37
Hollins Road
Darwin
Lancs BB3 OBG

Blue Circle Industries plc,
Northfleet Business Unit
Customer Services Office
The Shore
Northfleet
Kent DA11 9AN

British Gypsum Ltd
Technical Service Department
East Leake
Loughborough
Leics LE12 6JT

Gypsum Products Development Association
c/o KPMG
165 Queen Victoria Street
London EC4V 4DD

Health and Safety Executive's Information Centre
Broad Lane
Sheffield S3 7HQ
(http://www.open.gov.uk/hse/hsehome.htm)

Jesmonite Technologies Ltd
The Old School
Stanton Lacy
Ludlow
Shropshire SY8 2AE
(email: jesmonite@aol.com)

Knauf
PO Box 133
Sittingbourne
Kent ME10 3HW
(email: knauf@knauf.co.uk)

Lafarge Plasterboard Ltd
Marsh Lane
Easton-in-Gordano
Bristol BS20 0NF

Llewellyn Ryland Ltd
Haden Street
Birmingham B12 9DB
(email: sales@llewellyn-ryland.co.uk)

Snowcem
Snowcem House
Therapia Lane
Croydon
Surrey CR9 4EY
(email: snowcem@wallmaster.c.uk)

APPENDIX 3

BACKGROUNDS

Table A.1   British Gypsum finishing gypsum plasters

Background/lining	Preferred finish	Alternative finish
Plasterboard	Thistle Board Finish	Thistle Multi-Finish
Carlite Browning	Carlite Finish	Thistle Multi-Finish
Carlite Tough Coat	Carlite Finish	Thistle Multi-Finish
Carlite Bonding Coat	Carlite Finish	Thistle Multi-Finish
Thistle Hardwall	Carlite Finish	Thistle Multi-Finish
Thistle Dri-Coat	Thistle Board Finish	Thistle Multi-Finish
Thistle X-Ray	Thistle Multi-Finish	Thistle Board Finish
Concrete (with or without PVA bonding agent)	Thistle Multi-Finish *or* Thistle Board Finish	
Sand/cement undercoats	Thistle Multi-Finish	Carlite Finish *or* Thistle Board Finish
Backgrounds (eg glazed) treated with PVA	Thistle Multi-Finish *or* Thistle Board Finish	

Appendix 3

Table A.2 Knauf plaster selector

Plaster Selector ———— Select System and Product by Performance ———→

| | Select Product by Background Condition ↓ | * Betokontakt must be used
** Betokontakt may be used after background assessment |

		System Type		TWO COAT
		Key Benefit	General Purpose Undercoat	Bonding Undercoat
		Product Name	Knauf GreyBand	Knauf GreyBand Bonding
		Coverage per 1000 kg	130–150 m²	100–130 m²
		Setting Time	1½–2 hrs	1½–2 hrs
Backgrounds	**Suction**	**Key**	**Application Thickness**	
Glazed bricks and tiled surfaces	Very Low*	Very Poor		11 mm
Closed textured or smooth lightweight aggregate concrete blocks	Low**	Poor		11 mm
Clay engineering brickwork dense concrete	Low**	Poor		8–11 mm
Restoration and renovation clay brickwork, blockwork	Low to Moderate**	Variable		
Dense clay brickwork (other than Engineering brickwork), calcium silicate brickwork, concrete blockwork and concrete brickwork	Low to Moderate**	Variable	11 mm	11 mm
Plasterboard – lath, baseboard, ivory and grey surfaces	Low	Moderate		8 mm
Expanded plastic boards	Low	Moderate		
Autoclaved aerated concrete slabs and blockwork	Variable	Moderate to Good	11 mm	11 mm
Sand and cement undercoats	Moderate to high	Moderate		
No fines concrete	Low**	Good	11 mm	8–11 mm
Woodwool cement slabs	Low	Good	11 mm	11 mm
Metal lathing	Low	Good	11 mm	11 mm
Open texture lightweight aggregate concrete blocks	Variable	Moderate to Good	11 mm	11 mm
Normal clay brickwork and blockwork	Moderate to high	Good	11 mm	11 mm
Knauf GreyBand	Moderate to high	Good		
Knauf GreyBand Bonding	Moderate to high	Good		
Knauf BlackBand	Moderate to high	Good		

Table A.2 *Contd.*

	MULTI FINISH	BOARD FINISH	ONE COAT	
Impact Resistant Undercoat	Multi Finish Plaster	Plasterboard Finishing Plaster	Hand Applied One Coat Plaster	Machine Applied One Coat Plaster
Knauf BlackBand	Knauf MultiCover	Knauf Universal Board Finish	Knauf GoldBand	Knauf Projection
90–150 m^2	550–700 m^2	550–700 m^2	90–220 m^2	70–200 m^2
1½–2 hrs	1½–2 hrs	1½–2 hrs	1½–2 hrs	2–3 hrs
Application Thickness				
11 mm			5–20 mm	5–20 mm
8–11 mm			5–20 mm	5 20 mm
11 mm			5–20 mm	5–20 mm
	2–5 mm	2–5 mm		
	2–5 mm	2–5 mm		
11 mm			5–20 mm	5–20 mm
	2–5 mm	2–5 mm		
8–11 mm			5–20 mm	5–20 mm
11 mm			5–20 mm	5–20 mm
11 mm			5–20 mm	5–20 mm
11 mm			5–20 mm	5–20 mm
11 mm			5–20 mm	5–20 mm
	2 mm			
	2 mm			
	2 mm			

Table A.3   Plasterboard types

Types	Finishing	Edge	Thickness	Widths	Lengths
Wallboard	Tape & fill or plastering	Tapered Square	9.5 12.5 15.0	900 1200	1800, 2350, 2400, 2500, 2700, 3000, 3300, 3600
Vapour check wallboard*	Tape & fill or plastering	Tapered Square	9.5 12.5 15.0	900 1200	1800, 2350, 2400, 2500, 2700, 3000, 3300, 3600
Plank	Tape & fill or plastering	Tapered Square	19.0	600	2350, 2400, 2700, 3000
Baseboard	Plastering	Square	9.5	900	1220
Vapour-check baseboard*	Plastering	Square	9.5	900	1220
Lath	Plastering	Round	9.5 12.5	400 600	1220
High impact wallboard	Tape & fill or plastering	Tapered	12.5	900	2350 2400
Firecheck board*	Tape & fill or plastering	Tapered Square edge	12.5 15.0	900 1200	1800, 2400, 2700, 3000, 3600
Moisture resistant board*	Tape & fill or plastering	Tapered	9.5 12.5 15.0	900 1200	2350, 2400, 2700, 3000
Fire & moisture check*	Tape & fill or plastering	Tapered	12.5 15.0	1200	2400 3000
Impact resistant plasterboard*	Tape & fill or plastering	Tapered	12.5	1200	2400, 2700, 3000
Acoustic plasterboard*	Tape & fill or plastering	Tapered	12.5 15.0	1200	2400, 2700, 3000
Contour wallboard			6.0	1200	2400

Thermalcheck wallboards are available with expanded polystyrene and with phenolic foam bonded to the back to provide increased thermal insulation. The various types are between 22 and 55 mm thick.
* Vapour check wallboard and vapour check baseboard have a metallised foil on the reverse side. Firecheck board contains glass fibre and vermiculite additives. Moisture resistant board contains silicon additives. Impact resistant plasterboard contains glass fibre and vermiculite additives and is faced with heavy-duty paper. Acoustic plasterboard has a high density core.

Table A.4 Plasterboard fixings. Fixing detailed in BS 5492: 1990 Code of Practice for Internal Plastering

Board	Vertical (Studs or furring)	Horizontal (Joists or furring)	Nails (Rustproof 7 mm heads)	Nogging supports	Scrim
9.5 Gypsum lath	450	400	30 mm 4 across width	At edge of ceiling only	Angles only
9.5 Gypsum baseboard	450	400	30 mm 125 mm centres	At edge of ceiling only	Yes
12.5 Gypsum lath	600	450	40 mm 125 mm centres	At edge of ceiling only	Angles only
12.5 Gypsum baseboard	600	450	40 mm 125 mm centres	At edge of ceiling only	Yes
19 Gypsum plank	800	750	50 mm 5 across width	At edge of ceiling only	Yes
Wallboard	Follow manufacturer's instructions				

APPENDIX 4

JESMONITE COMPOSITES

There are four types of Jesmonite:

(1) *Jesmonite acrylic composite*: A general purpose compound for pouring and laminating.
(2) *Jesmonite hard composite*: A hard pouring and laminating compound, having high compressive strength with low expansion and shrinkage characteristics.
(3) *Jesmonite engineering composite*: An alumina-based compound suitable for working temperatures up to 250°C, having a compressive strength of 40MPA with low shrinkage and thermal movement.
(4) *Jesmonite soft composite*: A soft material suitable for carving. Suitable for patterns and original models.

Basic Jesmonite is supplied as a two-part system consisting of a liquid acrylic resin and a calcium sulphate powder.

Surface finishes, in-mould

- The gel coat may be pigmented.
- The gel coat can have stone powders incorporated. Newly demoulded casts can be washed off for maximum effect.
- The manufacturer states that leaving on the resin-rich ex-mould surface forms a protection against water absorption and UV degradation, with smooth and pigmented finishes.
- The gel coat can have metal powders incorporated.

Surface finishes, ex-mould

- There is a Jesmonite oxidising agent that may be applied to a metal-gel-coated cast.
- Jesmonite acrylic sealer may be applied.
- Wax polishes may be applied to items to be used internally.
- Jesmonite acrylic and metal powder coatings may be applied and burnished or patinated.

Control agents

- Jesmonite retarder.
- Jesmonite accelerator.

Fillers

Heavy and light weights for bulking to control viscosity and/or reduce costs.

Moulds

All standard mould materials and types as for fibrous plastering.

Sealing

Mould materials that require sealing are as for fibrous plastering and Jesmonite sealer is used.

Release agents

As in fibrous plastering, the mould materials requiring a release agent are the rigid ones. Soft waxes can be used – refer to Manufacturer's information.

Mixing

Materials are proportioned by weight and are mixed with high-shear blades.

Production

Pouring: With or without a gel coat, the standard mix may be poured into one-piece open moulds or into complex inner and outer piece moulds.

Laminating: A gel coat is applied. A second gel coat is required on vertical surfaces; this may contain 1–2% 6 mm chopped glass fibre.

Laminate: The gel coat is ready when it 'does not come away on finger' but is still soft. The standard mix is used with two layers of glass to give 5–6 mm thickness.

(1) A thick coat of mix is applied to the gel coat.
(2) One layer of glass fibre is laid on and worked on to the gel coat, using a brush.
(3) More mix containing 1–2% 6 mm chopped glass fibre is applied as a sandwich layer between the two layers of glass fabric.
(4) A second layer of glass fabric is applied.
(5) This layer is incorporated by brushing on a further mix without chopped glass fibre.
(7) Finally, a GRP laminating roller may be used to eliminate trapped air and achieve maximum compaction.

Reinforcements

Glass fibre

(1) Chopped strand 6 mm and 13 mm E glass fibre.
(2) A specially developed stitched glass fabric in two forms: Stitched quadaxial fabric and stitched biaxial fabric. The quadaxial fabric is a weave with two yarns at 90° to each other and two others at 45° angles. All are stitched at the intersection in such a way that they are able to move. This allows the weave to follow three-dimensional contours. The biaxial fabric has only two strands at 90° and is suitable for flat areas, producing a laminate that is less strong.
(3) Unifilo continuous strand mat.

Polypropylene

3-mm chopped fibres incorporated in the gel coat at 0.5–1% shows little or no trace.

For full information on Jesmonite systems and methods of use contact the manufacturers. See the Useful Addresses Appendix on p. 288.

J-FOAM

J-Foam is a flexible polyurethane foam casting compound in a two-part mix:

> Part A (grey) – polyether polyols
> Part B (clear brown) – diphenylmethane-4, 4-diisocyanate

The proportions used are around equal quantities by weight. Shortly after mixing, the resultant runny mix may be poured into open or closed inner and outer moulds (in the manner of pouring a case mould). There is an exothermic reaction and the mix foams and sets. The cooler surface of the mould causes a smooth, self-forming skin to appear on the surface of the cast. A quantity of the mix may be brushed on to the mould immediately before pouring to eliminate air holes and form a thicker skin.

With piece moulds, a portion of the mix may be brushed and/or poured on to each piece and the mould assembled, as long as provision has been made for the mould to vent ullage air and surplus expanding material.

According to quantity, as with all exothermic materials, the mix starts to foam minutes after mixing and is ready for demoulding in about 20 minutes, again depending on volume. The cast can be reinforced with soft metal (aluminium) sections. Spray waxes are used as release agents.

For full information consult the supplier's instructions. See also the section in Appendix 1 on polyurethane foams.